Seasonal Energy Storage with Power-to-Methane Technology

Seasonal Energy Storage with Power-to-Methane Technology

Editor

Attila R. Imre

MDPI • Basel • Beijing • Wuhan • Barcelona • Belgrade • Manchester • Tokyo • Cluj • Tianjin

Editor
Attila R. Imre
Budapest University of
Technology and Economics
Hungary

Editorial Office
MDPI
St. Alban-Anlage 66
4052 Basel, Switzerland

This is a reprint of articles from the Special Issue published online in the open access journal *Energies* (ISSN 1996-1073) (available at: https://www.mdpi.com/journal/energies/special_issues/sespmt).

For citation purposes, cite each article independently as indicated on the article page online and as indicated below:

LastName, A.A.; LastName, B.B.; LastName, C.C. Article Title. *Journal Name* **Year**, *Volume Number*, Page Range.

ISBN 978-3-0365-4889-0 (Hbk)
ISBN 978-3-0365-4890-6 (PDF)

© 2022 by the authors. Articles in this book are Open Access and distributed under the Creative Commons Attribution (CC BY) license, which allows users to download, copy and build upon published articles, as long as the author and publisher are properly credited, which ensures maximum dissemination and a wider impact of our publications.

The book as a whole is distributed by MDPI under the terms and conditions of the Creative Commons license CC BY-NC-ND.

Contents

About the Editor . . . . **vii**

Attila R. Imre
Seasonal Energy Storage with Power-to-Methane Technology
Reprinted from: *Energies* **2022**, *15*, 712, doi:10.3390/en15030712 . . . . **1**

Zoltán Csedő, Botond Sinóros-Szabó and Máté Zavarkó
Seasonal Energy Storage Potential Assessment of WWTPs with Power-to-Methane Technology
Reprinted from: *Energies* **2020**, *13*, 4973, doi:10.3390/en13184973 . . . . **3**

Gábor Pintér
The Potential Role of Power-to-Gas Technology Connected to Photovoltaic Power Plants in the Visegrad Countries—A Case Study
Reprinted from: *Energies* **2020**, *13*, 6408, doi:10.3390/en13236408 . . . . **25**

Gábor Pörzse, Zoltán Csedő and Máté Zavarkó
Disruption Potential Assessment of the Power-to-Methane Technology
Reprinted from: *Energies* **2021**, *14*, 2297, doi:10.3390/en14082297 . . . . **39**

Kristóf Kummer and Attila R. Imre
Seasonal and Multi-Seasonal Energy Storage by Power-to-Methane Technology
Reprinted from: *Energies* **2021**, *14*, 3265, doi:10.3390/en14113265 . . . . **61**

Máté Zavarkó, Attila R. Imre, Gábor Pörzse and Zoltán Csedő
Past, Present and Near Future: An Overview of Closed, Running and Planned Biomethanation Facilities in Europe
Reprinted from: *Energies* **2021**, *14*, 5591, doi:10.3390/en14185591 . . . . **75**

Michael Sterner and Michael Specht
Power-to-Gas and Power-to-X—The History and Results of Developing a New Storage Concept
Reprinted from: *Energies* **2021**, *14*, 6594, doi:10.3390/en14206594 . . . . **101**

Márk Szuhaj, Roland Wirth, Zoltán Bagi, Gergely Maróti, Gábor Rákhely and Kornél L. Kovács
Development of Stable Mixed Microbiota for High Yield Power to Methane Conversion
Reprinted from: *Energies* **2021**, *14*, 7336, doi:10.3390/en14217336 . . . . **119**

About the Editor

Attila R. Imre

Attila R. Imre received his Master degree in Physics from Eötvös University, Budapest, in 1990; four years later, he obtained a PhD from the same institution. He spent several years in various US and German universities, such as the University of Tennesee, Johannes Gutenberg Universität Mainz and Universität zu Köln. Meanwhile, in Hungary, he had various positions in the Atomic Energy Research Institute. Later, this institute became part of the newly formed Centre for Energy Research, where, recently, A. R. Imre has become a Scientific Advisor; he is also the Head of the Scientific Council of the research centre. After obtaining the Doctor of Science title from the Hungarian Academy of Science in 2015, he was appointed as Full Professor of Energy Engineering in the Budapest University of Technology and Economics; presently, he is the Head of the Energy Engineering Department. His present research topics cover various fields of energy production and storage, including the utilization of low-temperature heat sources (such as geothermal, solar and industrial waste heat) by ORC and similar methods and the application of power-to-gas technologies in energy storage. He has authored around 100 papers in various international journals.

Editorial

Seasonal Energy Storage with Power-to-Methane Technology

Attila R. Imre [1,2]

1 Faculty of Mechanical Engineering, Department of Energy Engineering, Budapest University of Technology and Economics, Műegyetem rkp. 3, H-1111 Budapest, Hungary; imreattila@energia.bme.hu
2 Department of Thermohydraulics, Centre for Energy Research, POB. 49, H-1525 Budapest, Hungary

Citation: Imre, A.R. Seasonal Energy Storage with Power-to-Methane Technology. *Energies* 2022, *15*, 712. https://doi.org/10.3390/en15030712

Received: 4 January 2022
Accepted: 16 January 2022
Published: 19 January 2022

Publisher's Note: MDPI stays neutral with regard to jurisdictional claims in published maps and institutional affiliations.

Copyright: © 2022 by the author. Licensee MDPI, Basel, Switzerland. This article is an open access article distributed under the terms and conditions of the Creative Commons Attribution (CC BY) license (https:// creativecommons.org/licenses/by/ 4.0/).

To have a sustainable society, the need to use renewable sources to produce electricity is inevitable. Due to the weather dependence of some of these sources (wind, solar), utility-scale energy storage has to be used. These fluctuations range from minutes (passing cloud) to whole seasons (Winter/Summer solar availability). Short-time storage can be solved (at least theoretically) with batteries. However, seasonal storage—due to the amount of storable energy and the self-discharging of some storage methods—is still a challenge to be solved in the near future.

Recently, novel methods are available among the classical long-term storage technologies (such as pumped hydro storage). Batteries are becoming better and better with less self-discharge and bigger energy density; therefore, they can be used for seasonal storage, although they cannot cover the total need. Therefore, Power-to-Gas methods (mainly Power-to-Hydrogen, P2H, and Power-to-Methane, P2M) play a bigger and bigger role in the storage mix. In these methods, surplus electricity is used to electrolyze water and produce hydrogen; this can then be stored and used later to recover electricity. Due to technical difficulties related to long-term hydrogen storage, alternative methods (such as Power-to-Methane or Power-to-Ammonia) can also be attractive solutions.

In Power-to-Methane technology, the hydrogen—with added carbon dioxide—can be turned to methane through chemical or biochemical methods. The methane can be stored and used later to recover electricity. Comparing the P2H and P2M methods, the energy recovery ratio is better for P2H; nonetheless, loss-free storage and recovery needs special equipment. By contrast, for P2M—being the produced methane SNG, i.e., synthetic natural gas—existing gas-storage facilities can be used for storage, and recovery can be achieved through the existing mature methods (such as gas engines). Although electricity recovery is associated with carbon dioxide emission, the amount of emitted CO_2 is equal to the one used for the synthesis; therefore, this technology can also be considered carbon-free.

There are two well-established ways for hydrogen-to-methane conversion: chemical and biochemical. The chemical way (the so-called Sabatier reaction) is fast and efficient, but it is a high-pressure and high-temperature reaction, which can be performed in special equipment; additionally, it might require hardly accessible metals for catalysis. Although sometimes it can be slower, the biochemical method is a low-temperature and low-pressure method utilizing microorganisms; some can be found even in biogas facilities. An additional advantage for the biochemical method is that it can be used on CH_4/CO_2 mixtures, i.e., it can enrich biogas to SNG.

This Special Issue is dedicated to biochemical Power-to-Methane technology. P2M technology is now on the verge of full-scale industrial use; therefore, a Special Issue dedicated to this method is very timely. The topics covered here range from basic biochemical research through comparison of various storage methods to complete energy storage solutions.

The increasing percentage of weather-dependent renewables in the energy mix forced researchers to find novel solutions for energy storage to fulfil the need for temporal balancing. In their paper, Sterner and Spechts [1] portrayed the 30-year-long history that led to Power-to-Everything (including Power-to-Methane and other Power-to-Fuel) technologies.

Szuhaj et al. [2] described the development of stable mixed microbiota for high yield Power-to-Methane conversion. This is a significant result because, with this method, it is unnecessary to use special strains for biomethanation; still, it was possible to enrich the initial biogas up to 95% of CH_4.

Kummer and Imre [3] compared other methods available for seasonal energy storage. They developed a simple function to help the ranking of various energy storage methods using their combined losses during unloaded and loaded time intervals.

P2M is not only a methane-producing technology; it has unique attributes because of renewable gas production, high-capacity grid balancing, and combined long-term energy storage with decarbonization, representing substantial innovation. Due to these points, the expected impact of P2M technology will be remarkable; the potentials hidden in this technology were outlined in the paper of Pörzse et al. [4].

For historical reasons, Visegrad countries (Czech Republic, Slovakia, Poland, and Hungary) have high-capacity gas storage and distribution networks, primarily built in the 1960s to the 1980s. Due to these capacities, P2M technology is a very attractive seasonal storage method in these countries because the produced methane can be stored and transported in their existing gas network. A case study to use P2M technology on the V4 countries in the regulation of Photovoltaic Power Plants were given by Pintér [5].

Concerning applicability, the biochemical P2M method can be appealing for countries with existing biogas production facilities. The paper of Csedő et al. [6] analyzes the financial side of the application of P2M technology in wastewater treatment plants as a seasonal energy storage facility, using Hungarian data.

Finally, Zavarkó et al. [7] reviewed the status of the technology by giving a critical review of closed, running, and planned biomethanation facilities in Europe. According to their results, future projects should have an integrative view of (chemical) hydrogen storage and utilization with carbon capture and utilization (HSU&CCU). In this way, the enhanced decarbonization potential would increase sectoral competitiveness.

We believe that biological Power-to-Methane technology—especially combined with biogas refinement—will be a significant player in the energy storage market within less than a decade. The ease of storage and use of methane as well as the effective carbon-freeness can make it a competitor for batteries or hydrogen-based storage, especially for storage times exceeding several months.

Funding: This research received no external funding.

Conflicts of Interest: The authors declare no conflict of interest.

References

1. Sterner, M.; Specht, M. Power-to-Gas and Power-to-X—The History and Results of Developing a New Storage Concept. *Energies* **2021**, *14*, 6594. [CrossRef]
2. Szuhaj, M.; Wirth, R.; Bagi, Z.; Maróti, G.; Rákhely, G.; Kovács, K.L. Development of Stable Mixed Microbiota for High Yield Power to Methane Conversion. *Energies* **2021**, *14*, 7336. [CrossRef]
3. Kummer, K.; Imre, A.R. Seasonal and Multi-Seasonal Energy Storage by Power-to-Methane Technology. *Energies* **2021**, *14*, 3265. [CrossRef]
4. Pörzse, G.; Csedő, Z.; Zavarkó, M. Disruption Potential Assessment of the Power-to-Methane Technology. *Energies* **2021**, *14*, 2297. [CrossRef]
5. Pintér, G. The Potential Role of Power-to-Gas Technology Connected to Photovoltaic Power Plants in the Visegrad Countries—A Case Study. *Energies* **2020**, *13*, 6408. [CrossRef]
6. Csedő, Z.; Sinóros-Szabó, B.; Zavarkó, M. Seasonal Energy Storage Potential Assessment of WWTPs with Power-to-Methane Technology. *Energies* **2020**, *13*, 4973. [CrossRef]
7. Zavarkó, M.; Imre, A.R.; Pörzse, G.; Csedő, Z. Past, Present and Near Future: An Overview of Closed, Running and Planned Biomethanation Facilities in Europe. *Energies* **2021**, *14*, 5591. [CrossRef]

Article

Seasonal Energy Storage Potential Assessment of WWTPs with Power-to-Methane Technology

Zoltán Csedő [1,2], Botond Sinóros-Szabó [1] and Máté Zavarkó [1,2,*]

[1] Power-to-Gas Hungary Kft, 5000 Szolnok, Hungary; csedo@p2g.hu (Z.C.); sinoros@p2g.hu (B.S.-S.)
[2] Department of Management and Organization, Corvinus University of Budapest, 1093 Budapest, Hungary
* Correspondence: mate.zavarko@uni-corvinus.hu

Received: 29 August 2020; Accepted: 18 September 2020; Published: 22 September 2020

Abstract: Power-to-methane technology (P2M) deployment at wastewater treatment plants (WWTPs) for seasonal energy storage might land on the agenda of decision-makers across EU countries, since large WWTPs produce a notable volume of biogas that could be injected into the natural gas grid with remarkable storage capacities. Because of the recent rapid increase of local photovoltaics (PV), it is essential to explore the role of WWTPs in energy storage and the conditions under which this potential can be realized. This study integrates a techno-economic assessment of P2M technology with commercial/investment attractiveness of seasonal energy storage at large WWTPs. Findings show that a standardized 1 MW_{el} P2M technology would fit with most potential sites. This is in line with the current technology readiness level of P2M, but increasing electricity prices and limited financial resources of WWTPs would decrease the commercial attractiveness of P2M technology deployment. Based on a Hungarian case study, public funding, biomethane feed-in tariff and minimized or compensated surplus electricity sourcing costs are essential to realize the energy storage potential at WWTPs.

Keywords: seasonal energy storage; power-to-methane; wastewater treatment plants; techno-economic assessment

1. Introduction

There is broad consensus within the power-to-gas (P2G) literature, especially in the power-to-methane (P2M) literature, as well as among industry actors that wastewater treatment plants (WWTPs) could play a significant role in scaling up P2G technology by ensuring key input factors, mainly efficiently useable carbon-dioxide sources in the produced biogas [1]. Meanwhile, a notable volume of previous research has shown several technical, and techno-economic challenges of the P2M technology [2], and recent research has also pointed out that a supportive regulatory environment is essential to further develop and scale up the P2M technology [3]. As the EU must significantly increase the PV installation rate to reach a carbon-neutral electricity supply by 2050 [4], and considering the integration challenges of the renewable energy to the grid [5], it is becoming a key priority for decision-makers to also focus on concrete opportunities and limitations of seasonal energy storage that could be realized with P2M technology deployment at WWTPs.

While the promising role of the P2G technology in the energy sector has been argued comprehensively in recent years (e.g., from the aspect of long-term energy storage [2], system analysis [6] or technological and economic factors [7]), researchers have also started to focus on the role of WWTPs with respect to different aspects of renewable energy transition and power-to-X technologies. Schäfer et al. [8] pointed out that WWTPs have notable synergy potential in sector coupling, for example, hydrogen and methane can be produced at WWTPs (with P2G technologies), and the oxygen (as the byproduct of the electrolysis) can be used to enhance purification processes. Gretzschel et al. [9] focused on power-to-hydrogen (P2H) technology and the elimination of organic micropollutants at

WWTPs, considering the possibility of offering system service, as well: automatic frequency restoration reserve (aFRR), which can provide short-term flexibility for network operators. Ceballos-Escalera et al. [10] examined the energy storage attributes of a prototype with a bioelectrochemical system for electromethanogenesis (EMG-BES) at a WWTP, which is an emerging technology in the P2M segment besides chemical and biological methanation. They also showed the future potential of the interconnectedness of renewable energy overproduction, biomethane production, and wastewater treatment. WWTP functions regarding sustainability are, however, researched in terms of other aspects, as well, considering that they also play a significant role in nutrient recovery, where new practices have been suggested [11], and have also been designed [12] for environmentally and economically more viable clarification and treatment technologies.

In this paper, the authors make a step forward on the route outlined by these previous researchers, using Hungary as a case study and focusing on biological methanation technology. Its technology readiness level makes it possible to plan grid-scale implementations, even in the short term [13]. These opportunities are paved by the theoretical synergies between biological methanation and WWTPs mentioned above, as well as empirical data of:

(1) the innovative lab-scale P2G prototype with biological methanation developed by Power-to-Gas Hungary Kft. in cooperation with Electrochaea GmbH (the developer of the 1 MW_{el} P2G facility with biomethanation, located in Avedøre, Denmark).

(2) large Hungarian WWTPs, from which the authors collected technical data to evaluate the implementation opportunities and limitations. The senior executives of these WWTPs provided valuable insights regarding the economic and technology incentives of commitment for grid-scale technology implementation projects.

Techno-economic assessments have already been conducted regarding P2G technologies with different methods and scopes in recent years. In terms of the return of the investment, for example, Ameli et al. [14] analyzed the role of different capacities of battery storage and P2G systems in Great Britain with the Combined Gas and Electricity Networks (CGEN) model. Addressing electricity balancing challenges, they concluded that the capital costs must reach £0.5 m/MW for P2G to justify the investment. As a comparative approach, Collet et al. [15] analyzed five different scenarios of biogas upgrading and P2G, pointing out that P2G technologies "are competitive with upgrading ones for an average electricity price equal to 38 EUR MW h^{-1} for direct methanation and separation by membranes" [p. 293]. In the case of production costs, Peters et al. [16] can be mentioned among others, who evaluated eight scenarios based on different combinations of H_2 and CO_2 sources and found methane costs in the range of 3.51–3.88 EUR/kg for P2G. Collet et al. and Peters et al. complemented their techno-economic analyses with ecological and environmental aspects, focusing on greenhouse gas (GHG) emissions, as well. P2M is important, but is not the only means of decarbonization in the case of waste management; for example, the latest techno-economic analyses show increasing economic and ecological viability regarding biochar farming [17], agricultural waste management [18] and solid biofuels [19], as well.

As detailed above, there are several approaches to perform a techno-economic assessment of the P2G and waste management technology [20]. Inspired by these studies, the authors also emphasize the economic aspects besides the technical parameters, based on which the seasonal energy storage potential can be calculated at large Hungarian WWTPs. The novelty of this paper is that it aims to open up new perspectives in the techno-economic assessment of P2M technology by:

(1) narrowing its focus to individual WWTPs in the first step and carrying out in-depth analysis regarding not only techno-economic, but also commercial/investment questions as complementary viewpoints (in addition to the important and frequently assessed environmental impacts). Economic and commercial aspects are differentiated, as the former considers general interrelations of technical data, costs, revenues, return on investment; while the latter incorporates WWTP-specific infrastructure, strategic management and investment related viewpoints of WWTPs as organizations, as well.

(2) extending the focus to a national-level assessment, based on specific empirical data, as well as evaluating the seasonal energy storage and practical implementation opportunities and limitations of the P2M technology with integrated commercial/investment expectations of stakeholders.

Consequently, the research questions are the following:

(1) What is total seasonal energy storage potential with P2M at large WWTPs in Hungary?

(2) What are the economic conditions under which WWTPs are financially incited to participate in a grid-scale implementation of the biological methanation?

The research questions indicate that the authors aim to connect theory and practice, to explore the seasonal energy storage potential and also the practical key success factors under which this potential can be realized. The focus of this paper makes meaningful contributions to P2M research and industry that are beyond the specific geographical area:

(1) First, while numerous studies have drawn important conclusions about the "hard" factors of P2G technology development and implementation (such as levelized cost of energy, process design, cost optimization, life-cycle assessment) based on quantitative data [2], the authors combine quantitative and qualitative data collection to contribute to an overall understanding of P2M technology deployment opportunities and limitations at concrete future operators of P2M.

(2) Second, the techno-economic assessment with the complementary commercial/investment viewpoint (based on interviews and financial modeling) shows how WWTPs senior executives could be incited by changes of the regulatory environment to take the innovation-related and upscaling risks, as well. Figure 1 summarizes the research framework.

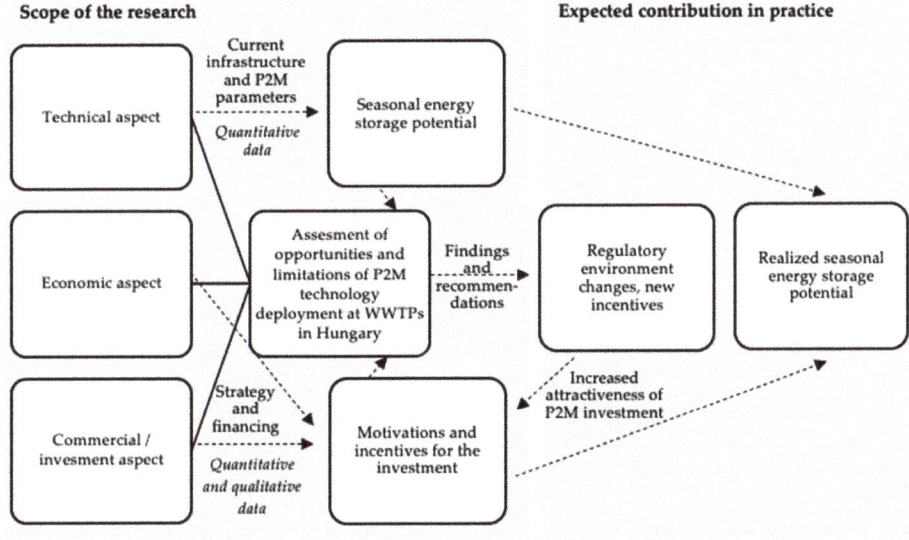

Figure 1. Research framework. The scope of the research incorporates the assessment of opportunities and limitations of P2M technology deployment at WWTPs in Hungary in terms of technical, economic, and commercial/investment aspects. The current WWTP infrastructure and P2M technology parameters determine the seasonal energy storage potential. Commercial and investment challenges of WWTPs and P2M business models determine motivations and incentives for such projects. Based on these findings, recommendations can be outlined for changes of the regulatory environment. The expected contribution of these recommendations is that new incentives could increase the attractiveness of P2M investments for WWTPs and allow them to realize the energy storage potential.

As the research framework suggests, based on a previous Hungarian P2G study [3], the specific research hypothesis is the following:

Economic, commercial, and investment aspects of P2M seasonal energy storage do not motivate WWTPs to act as future P2M operators, consequently, there is a need for change in the regulatory environment to incite them to realize their seasonal energy storage potential with P2M deployment.

There is a rapidly growing need for seasonal energy storage in the EU, especially in Hungary (where the national energy strategy also forecasts rapid growth of national PV capacities [21]), for which P2M would be a promising technology, but its grid-scale implementation has not happened yet. The objective of the research is to examine the P2M deployment opportunities and limitations at large WWTPs in Hungary and explore possible ways of realizing the seasonal energy storage potential of P2M technology. The main contribution of this techno-economic assessment is that it incorporates complementary commercial and investment attractiveness of seasonal energy storage by collecting and analyzing both quantitative and qualitative data as well. It shows the challenges of P2M technology deployment also from the aspect of future operators highlighting their motivation and strategic interests.

2. Materials and Methods

2.1. Technology Description

P2G is often called a "disruptive" technology, since it brings a new techno-socio-economic approach into the energy sector and redefines the scope of duties of each stakeholder (Ferrero, 2016). This disruptive process started in Hungary with the foundation of Power-to-Gas Hungary Kft., in 2016. The startup developed a lab-scale P2M prototype and has been operating it since April 2018. The prototype is a scaled-down operational unit with mass and energy flows in proportion to the commercial process of P2G, and also contains the complete basic unit operations to carry out research and development (R&D) in the field of P2G.

The planned P2M plants can produce a gas mixture that meets the requirements of natural gas standards. The applied process consists of three main steps.

(1) In the power-to-hydrogen (electrolysis) step, the plant would use surplus electricity from the electric grid [22] and produce hydrogen (with oxygen as a byproduct), in line with the chemical reaction below:

$$4 H_2O \text{ (l)} + e^- \rightarrow 4H_2 \text{ (g)} + 2O_2 \text{ (g)}, \quad \Delta H_r^0 = 285.5 \text{ kJ/mol} \tag{1}$$

In this research, polymer electrolyte membranes (PEMEC) electrolysis is applied, which is preferred for seasonal energy storage (as it is applied also by Power-to-Gas Hungary Kft), mainly because of its high flexibility, fit to volatile renewable energy generation, and high technology-readiness level [23]. While hydrogen is going to be used in the next P2G step (methanation), oxygen generation can also be exploited at WWTPs; the efficiency of the aeration system can be increased by injection of oxygen into it [8].

(2) In the methanation step, the CO_2 content of the biogas (typically 30–50%) is converted to methane, carried out by basic reactions and mediated by the biocatalyst employing a unique set of enzymes [24]:

$$CO_2 + 4H_2 \rightarrow CH_4 + 2H_2O \tag{2}$$

In this research, a flexible biomethanation process is applied that is provided by an optimized strain of Archaea (Methanothermobacter thermautotrophicus), a proprietary biocatalyst, a robust, highly selective and efficient strain [25]. Unlike biogas upgrading [26], methane and carbon dioxide gas components are not separated in this process, and the biogas is injected to the continuous stirred-tank reactor along with hydrogen. Mass-flow rates are set to maintain the stoichiometric ratio of hydrogen and carbon dioxide (increased, 4.1:1 in practice because of the 23 times lower dissolution of hydrogen than carbon dioxide in water).

(3) In the injection step, the product gas in which the guaranteed purity of methane is more than 97% is injected into the natural gas grid after a polishing process (segregation of hydrogen gas compound, removal of water vapor, cooling).

The evaluated total efficiency of the P2M plant (η_{P2G}) is calculated as follows:

$$\eta_{P2G} = \eta_{el} \cdot \eta_{meth} = \left(\frac{\dot{V}_{H2} \cdot HHV_{H2}}{3.6 \cdot P_{el}} \cdot \frac{\dot{V}_{wpg} \cdot HHV_{wpg}}{\dot{V}_{H2} \cdot HHV_{H2}} \right) \cdot 100 = 27.7 \cdot \frac{\dot{V}_{wpg} \cdot HHV_{wpg}}{P_{el}} \ [\%] \qquad (3)$$

where:
$\dot{V}_{H2}$ – Hydrogen gas volumetric flow $\left(\frac{Nm^3}{h}\right)$
HHV_{H2} – Hydrogen gas higher heating value $\left(\frac{MJ}{Nm^3}\right)$
P_{el} – power output of electrolyzer units [kW_{el}]
$\dot{V}_{wpg}$ – wet product (effluent) gas volumetric flow $\left(\frac{Nm^3}{h}\right)$
HHV_{wpg} – wet product (effluent) gas higher heating value $\left(\frac{MJ}{Nm^3}\right)$

After substituting the correspondent values into the equation, the total P2M plant efficiency is in the range of 55–60%.

2.2. WWTPs in Hungary

WWTPs in Hungary are units of regional or municipal waterworks, typically owned by municipals responsible for water supply, wastewater drainage, and treatment. There were 826 WWTPs in Hungary in 2016, ca. 96% of which were under 100,000 PE (Population Equivalent). Considering the goal of grid-scale P2M technology implementation and its complex infrastructural and input conditions [3], the 28 WWTPs above 100,000 PE could be relevant for this research. Not every WWTP with large PE produces biogas, however (for example, the authors found that only 13 WWTPs have biogas plants from the 19 WWTPs of Hungary's county seats), but there are other WWTPs at non-county seats which also have biogas. In sum, there are around 20 WWTPs with favorable infrastructure that produce biogas in Hungary. In 2016, the calorific value of biogas was 897,066,000 MJ/year on the national level [27].

2.3. Data Collection

The authors analyzed the implementation potential of the innovative and efficient biomethanation technology of Power-to-Gas Hungary Kft. At different sites, technical data was collected from large Hungarian WWTPs, and several interviews were carried out at the level of experts and senior executives, as well. The authors were able to collect data from seven WWTPs from four different regions of Hungary, which is in line with the decentralization trends of the energy sector [28]. As all of the analyzed WWTPs were above 100,000 PE, this research represents the biggest cities of Hungary.

The data collection process contained at least four steps in every case:

(1) Pre-evaluation of the P2M technology relevancy with the Chief Technology Officer or the Technical Director (semi-structured interviews);

(2) In-depth presentation of the technology and exploration of the commercial opportunities with the Chief Executive Officer or the executive team (semi-structured interviews or focus group interviews);

(3) Collection of existing techno-economic data and documentation;

(4) On-site techno-economic data collection and consultation.

Table 1 shows the structure of the data collection. Because of confidentiality, specific financial data were provided only in terms of trends, or highlighting opportunities and challenges.

Table 1. Structure of data collection.

Data	Technical, Technological, Infrastructural	Economic, Commercial, Investment Related
General (Senior executive and director level)	<ul><li>Power supply from grid, current or planned PV capacity</li><li>Water supply</li><li>CO_2 input: % in the biogas, produced volume per h</li><li>The geographical area for the P2M plant and its local infrastructural connections (for example to the biogas plant or the WWTP)</li><li>Connection to the natural gas grid</li><li>Byproduct use potential (waste heat, oxygen)</li></ul>	<ul><li>Openness for technological innovations and collaborations</li><li>Financial situation</li><li>Current biogas use</li><li>Current or planned infrastructural developments, potential synergies with P2M</li></ul>
Specific (Director and expert/level)	<ul><li>Fermentation (e.g., temperature)</li><li>Raw biogas composition (e.g., sulfur)</li><li>Gas characteristics (e.g., gas flow, pressure)</li><li>Gas engines (e.g., type, electric and thermal power)</li><li>Power grid connection (e.g., voltage)</li><li>Natural gas grid connection (e.g., distance from the plant)</li><li>Water and wastewater (e.g., treatment technology)</li><li>Technological and infrastructural connections (e.g., current or possible use of waste heat)</li><li>Expansion potential (e.g., transport connections, geographical area).</li></ul>	<ul><li>Mobilizable capital for the investment</li><li>Current contracts defining energy costs</li><li>Current revenues produced or costs saved on biogas use</li></ul>

Moreover, the authors conducted interviews with technology suppliers, researchers, strategic and financial investors, and other stakeholders in the P2G inter-organizational innovation networks [3] as well, which helped to contextualize the former techno-economic analyses and the new data from WWTPs in Hungary.

2.4. Data Analyses

2.4.1. Applied Model for the Calculation of Seasonal Energy Storage Potential

The seasonal energy storage potential can be calculated on the basis of HHV of the total generated injected gas. The parameters of the injected gas mixture must meet the gas requirements set in Hungarian Standards [29] and Annex 13 of Implementing Regulation of Natural Gas Supply [30]. The most significant specifications to meet are

- Wobbe index: 45.66–54.76 MJ/m^3
- HHV: 31.00–45.28 MJ/m^3 (8.61–12.58 kWh/m^3)
- Hydrogen sulfide content: max. 20 mg/m^3
- Water vapor content: 0.17 g/m^3

Since the polished wet gas carbon dioxide concentration exceeds 97%, the higher heating value of the injected gas (HHV_{P2G}) is calculated as follows:

$$HHV_{P2G} = 0.97 \cdot HHV_{CH_4} = 0.97 \cdot 36.3 \, \frac{MJ}{Nm^3} = 35.21 \, \frac{MJ}{Nm^3} = 9.78 \, \frac{kWh}{Nm^3} \quad (4)$$

2.4.2. Applied Model for the Economic Analysis

The economic analysis is based on a single "average" WWTP case in the first step and extends the scope to the national level in the second. The authors built their financial calculations largely on the data and the analyses of the EU-funded STORE&GO project. This project was focused on three variations of P2G implementation since 2016, one of them with biological methanation [31].

The background driver of this economic analysis was the National Energy Strategy 2030 of Hungary, which aims towards the rapid growth of electricity generating units from photovoltaic sources (the planned installed capacity will exceed 6000 MW by 2030) [21]. This is indeed a favorable trend for the renewable transition. The literature has also pointed out, though, the challenges of surplus energy generation and the need for energy storage [32]. In this respect, the Hungarian natural gas grid would be appropriate for seasonal energy storage, with its 6,330,000,000 m^3 storage capacity [21].

The fundamental assumption of this economic analysis is that during the rapid growth of PV capacities in Hungary, the Hungarian feed-in tariff (FiT) system and its green premium [33], which provides higher electricity prices for renewable energy producers to incite more PV investments, negatively affects the P2M business model and its attractiveness for investors. As P2M technologies are key in energy storage [34], further regulatory changes and incentives are needed to avoid energy loss and network imbalance. There is a clear need for a system in which seasonal energy storage can be incited and realized but without impeding the further growth of PV capacity in the country. Figure 2 illustrates the background and the focus of the economic analysis of the study.

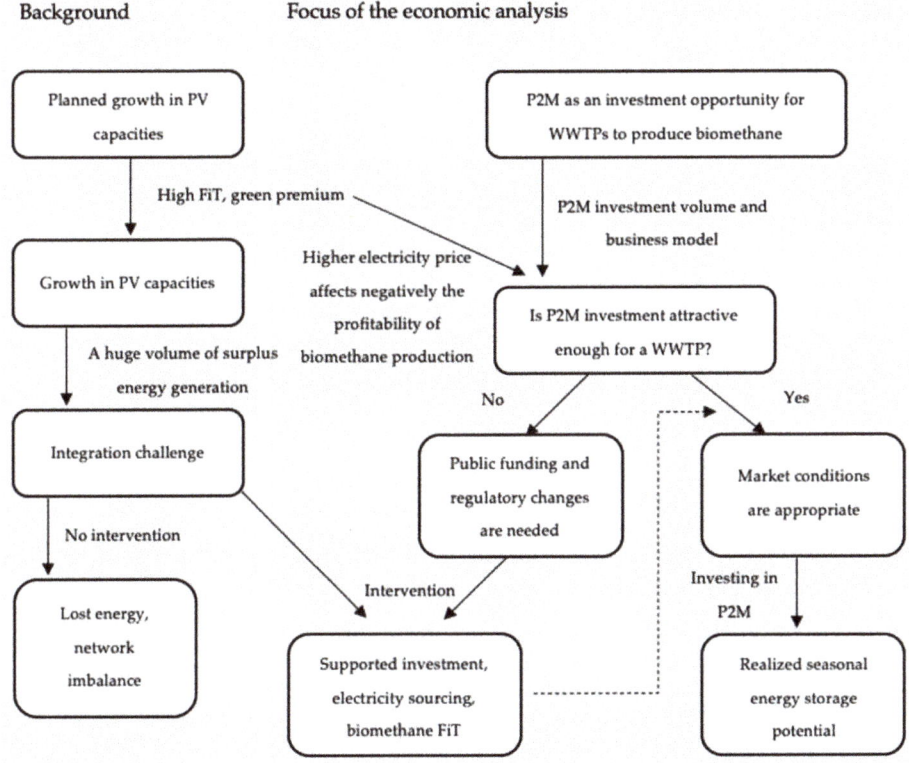

Figure 2. Model of economic analysis.

As previous studies have shown that electricity sourcing is the most determining factor of operating expenditures (OPEX) and the most economic benefit can be realized when the P2X plant is directly connected with PVs or wind turbines [35], one possible way to incite P2M investments could be to provide a framework in which P2G plants can use this surplus energy at well below market price, or if this sourcing were compensated as an acknowledged system service for flexibility or energy storage. This optimization of the price difference between the input (electricity) and the output (biomethane) on the cost side is relevant practically in ca. 1200 h per year [36] with respect to seasonal energy storage. Seasonal energy storage can be supported further on the revenue side, as biomethane FiT has been implemented in a few countries in Europe [37].

On the capital expenditures (CAPEX) side, EU-funded and state-funded projects can foster P2M investment, mostly with dominant research, development and innovation (R&D&I) focus, like in the case of the STORE&GO project [31]. These concepts are not far away from the approach of the National Energy Strategy 2030 of Hungary, because it plans to build a pilot, then a grid-scale P2G plant, a regulatory sandbox model, and a mandatory national purchasing system for biomethane [21].

Based on the technical parameters, the economic and business analysis explores whether current market conditions are attractive for WWTPs to invest in P2M technology or not. If not, the analysis identifies scenarios combining the incentive opportunities of the cost, the revenue, and the investment dimensions to meet the criteria of WWTP executives (identified during the interviews).

2.4.3. Qualitative Data Analysis

As mentioned before, the techno-economic assessment has a complementary commercial/investment viewpoint. Consequently, 21 interviews were conducted with senior executives and directors and analyzed using the coding technique of the grounded theory [38]. The approach of this data analysis method fits the functionalist research, as it provides a structured process (open coding, axial coding, selective coding) to build or fine-tune a theory (a general conclusion) [39] opposed to other (mostly interpretative) qualitative methods (e.g., qualitative content analysis [40]).

(a) To improve the validity, the authors continued the research even after the fourth and fifth cases, even though they did not obtain significantly new information compared to the previous ones (reached theoretical saturation [38]).

(b) To improve the reliability, validation of the pre-conclusions was asked about during the on-site consultations.

(c) To improve the generalizability, the interview questions were modified according to the conclusions of the previous case, testing whether these conclusions were valid in other contexts or not.

3. Results

3.1. Seasonal Energy Storage Potential

In this section, the authors present the theoretical seasonal energy storage potential at large Hungarian WWTPs; then they point out the difference between this theoretical potential and the practical potential, which is calculated based on their empirical data collection.

3.1.1. Storage Potential of an "Average" WWTP Case

As previously described, storage potential is evaluated by taking WWTPs exceeding 100,000 PE into consideration. Based on previous research, the biogas yield of an average sewage anaerobic digestion (AD) facility in Hungary reaches 0.04 m^3/day/PE [41]. The 20 WWTPs which are relevant in this study and exceeding 100,000 PE, have a combined PE value of 5,901,866. Based on the data above, the average size of Hungarian WWTPs that are relevant for P2G technology (C_{P2G}): $C_{P2G} = \frac{5,901,866 \text{ PE}}{20} = 295,093 \text{ PE}$

The average biogas yield of an average WWTP:

$$P_{P2G} = 0.04 \frac{Nm^3}{day \cdot PE} \cdot C_{P2G} = 0.04 \frac{Nm^3}{day \cdot PE} \cdot 295,093 \text{ PE} = 11,804 \frac{Nm^3}{day} \quad (5)$$

Presuming the methane ratio of the biogas yield is 0.55, the hourly volumetric carbon dioxide flow of an average WWTP is calculated by the equation below:

$$\dot{V}_{CO_2} = (1 - 0.55) \cdot \frac{P_{P2G}}{24} = 0.45 \cdot \frac{11,804}{24} = 221.2 \frac{Nm^3}{h} \quad (6)$$

The electrolyzer capacity of a P2G facility using biogas of an average WWTP is calculated with the presumption of the 4.7 kWh electrical energy demand for the yield of 1 Nm³ of biomethane is 4.7 kWh/Nm³:

$$P_{P2G} = \dot{V}_{H_2} \cdot 4.7 \frac{kWh}{Nm^3} = \dot{V}_{CO_2} \cdot 4.1 \cdot 4.7 \frac{kWh}{Nm^3} = 221.2 \frac{Nm^3}{h} \cdot 4.1 \cdot 4.7 \frac{kWh}{Nm^3} = 4263 \text{ kW} = 4.26 \text{ MW}_{el} \quad (7)$$

The other way of calculating P2G capacity for an average WWTP is by using the biogas volumetric flow rates burned in combined heat and power (CHP) units at WWTP sites. Kisari [42] defined regional WWTPs' onsite CHP capacity by analyzing 10 relevant biogas plants using biogas generated from anaerobic degradation of sewage slurry. In accordance with his research, the average built-in CHP capacity was 730 kW$_{el}$ (P_{CHP}). Sinoros [43] calculated the theoretical P2G potential with the focus on available regional bioethanol and biogas yield in Hungary. That research carried out conclusions on total biogas annual yield and considered no difference in the sources, particularly on WWTP biogas streams.

The calculation of P2G plant capacity on the basis of built-in CHP capacity of WWTPs:

$$P\prime_{P2G} = (\dot{V}\prime_{CO_2}) \cdot 4.1 \cdot 4.7 \frac{kWh}{Nm^3} = \left(\frac{P_{CHP} \cdot (1-0.55)}{\frac{\eta_{CHP}}{100} \cdot \left(\frac{100-r_s}{100}\right) \cdot HHV_{CH_4}} \right) \cdot 4.1 \cdot 4.7 \frac{kWh}{Nm^3} \quad (8)$$

$=$, where

r_s-AD plant electric self-consumption percentage—15%
η_{CHP}-CHP electric efficiency—35%
HHV_{CH_4}–Higher heating value of methane—10.3 kWh/Nm³

After executing the substitution, the calculated capacity is:

$$P\prime_{P2G} = \left(\frac{730 \text{ kW} \cdot (1-0.55)}{\frac{35}{100} \cdot \left(\frac{100-15}{100}\right) \cdot 10.3} \right) \cdot 4.1 \cdot 4.7 \frac{kWh}{Nm^3} = 107.2 \frac{Nm^3}{h} \cdot 4.1 \cdot 4.7 \frac{kWh}{Nm^3} = 2065 \text{ kW}_{el} = 2065 \text{ MW}_{el} \quad (9)$$

Although P_{P2G} is more than two times higher than $P\prime_{P2G}$, due to the constraints of site conditions the authors justified P2G potential at a lower value than $P\prime_{P2G}$. In accordance with the information collected onsite and all the datasets provided by WWTP site managers, a P2G plant with 1 MW$_{el}$ electrolyzer capacity could be fit to the WWTPs with the load exceeding 100,000 PE in general, because

(1) the methane content is usually higher (around 60–65%) than expected based on the literature, which is beneficial for biogas production but not for P2M, because there is less CO_2 (around 35–40%) to convert to biomethane;

(2) the raw biogas flow is around 130 Nm³/h on average at the empirically examined WWTPs, which slightly exceed 100,000 PE, but there are 9 WWTPs that are above even 250,000 PE (obviously they are still within the necessary scope for P2M deployment);

(3) there is some seasonality in the case of several WWTPs (e.g., at Lake Balaton) that affects biogas production, but the higher values are typically in the summer, which fits the seasonal energy storage concept.

3.1.2. Energy Storage Potential

According to Section 2.4.1, total seasonal energy storage potential can be calculated on the basis of the higher heating value of the injected gas. Based on the stoichiometry, the values of E and E' are calculated as follows:

$$E = \dot{V}_{CO_2} \cdot 9.78 \frac{kWh}{Nm^3} \cdot t_{OP} = 221.2 \frac{Nm^3}{h} \cdot 9.78 \frac{kWh}{Nm^3} \cdot 1200 \frac{h}{year} \\ = 2,596,004 \text{ kWh} \approx 2596 \text{ MWh} \quad (10)$$

$$E' = \dot{V'}_{CO_2} \cdot 9.78 \frac{kWh}{Nm^3} \cdot t_{OP} = 107.2 \frac{Nm^3}{h} \cdot 9.78 \frac{kWh}{Nm^3} \cdot 1200 \frac{h}{year} \\ = 1,258,099 \text{ kWh} \approx 1258 \text{ MWh} \quad (11)$$

Total theoretical seasonal energy storage potential of 20 WWTPs exceeding 100,000 PE is

$$E_{total} = E \cdot 20 = 2596 \text{ MWh} \cdot 20 = 51,920 \text{ MWh} \approx 51.9 \text{ GWh} \quad (12)$$

$$E'_{total} = E' \cdot 20 = 1258 \text{ MWh} \cdot 20 = 25,160 \text{ MWh} \approx 25.2 \text{ GWh} \quad (13)$$

Considering all the information collected in site visits, the practical seasonal energy storage potential of an average WWTP is

$$E_p = \dot{V}_{pCO_2} \cdot 9.78 \frac{kWh}{Nm^3} \cdot t_{OP} = 50 \frac{Nm^3}{h} \cdot 9.78 \frac{kWh}{Nm^3} \cdot 1200 \frac{h}{year} \\ = 586,800 \text{ kWh} \approx 587 \text{ MWh} \quad (14)$$

Total practical seasonal energy storage potential of 20 WWTPs exceeding 100,000 PE is

$$E_{ptotal} = 20 \cdot E_p = 20 \cdot 587 \text{ MWh} = 11,740 \text{ MWh} \approx 11.7 \text{ GWh} \quad (15)$$

3.2. Commercial and Investment Perspectives

3.2.1. Investment Volume, Operating Expenses, and Revenues

An important statement of the financial analyses of the STORE&GO project is that a high range of possible investment costs of electrolyzers and methanation systems can be seen in the literature [44]. The economies of scale are a determining factor of CAPEX [44]. The investment costs in this study are based on the calculations of van Leeuwen and Zauner [45] with minor modifications according to the technical infrastructure of the analyzed WWTPs and additional costs of a public-funded technology development projects. Interviewees also pointed out that one must take into account the costs of public grant/public financing-specific R&D and maintenance tasks, and furthermore, the needed software background supporting the P2M technology operations (not only the hardware and the physical infrastructure). Appendix A shows the basis of the CAPEX calculations.

(1) The specific investment cost of the PEM electrolyzer system is 1640 EUR/kW, which is the base case according to van Leeuwen and Zauner.

(2) In the case of the methanation system, a slightly higher CAPEX than the base case, 0.5 EUR/kW$_{el}$ is taken into account because of some high specific investment costs for biomethanation presented by Böhm et al. [44].

(3) There is an integrated "infrastructure" cost item, as well, because different kinds of infrastructure development are needed at the analyzed WWTPs (e.g., there is gas storage at a few WWTPs, or the new infrastructure for the use of the oxygen as a byproduct can be also relevant in this cost item).

(4) An additional 28% investment is needed for project development, planning, expert services, quality management, according to van Leeuwen and Zauner, and an additional 50% for public grant/public financing-specific R&D, software development, and maintenance tasks.

Based on the above, the CAPEX of a 1 MW_{el} P2M plant at an "average" WWTP is 5,696,000 EUR if the investment would be realized this year.

The deployment of even one P2M plant, however, could require even more than a year-long project planning, and 20 P2M plants cannot be deployed in one year. Consequently, the time horizon must be extended for the investment. Previous P2G research has shown that there is a significant cost reduction potential regarding investment costs because of experience curves and learning rates. Böhm et al. [44] calculated that PEMEC CAPEX will decrease from 1200 EUR/kW_{el} (2017) to 530 EUR/kW_{el} (2030), and biological methanation CAPEX will decrease from 600 EUR/kW_{SNG} (2017) to 360 EUR/kW_{SNG}. This means that CAPEX of these components will decrease by 55% and 45% in 13 years. As the authors in this research assume that P2M plants in question will be deployed between 2020 and 2030, in parallel with the planned growth of PV capacities in Hungary, some CAPEX reduction is needed based on the quoted estimation. Assuming even distribution of P2M deployment for the next 10 years, the year 2025 can be taken as the basis of the calculation, so the 1 MW_{el} P2M CAPEX for 2025 with PEMEC CAPEX can be decreased by 25% and the CAPEX of biomethanation system can be decreased by 20%. Consequently, the model calculates on the basis of the reduced, 4,806,000 EUR CAPEX.

In the economic analysis, this CAPEX was considered as a fixed component, while operating expenses and revenues were influenced by the costs of electricity sourcing (power grid fees) or its compensation, and biomethane price was considered as a variable contingent on potential regulatory changes. Appendix B shows the assumptions of the OPEX and revenue calculations. It is worth mention that besides biomethane, waste-heat could generate an important revenue stream at 55 EUR/MWh [46]; however, this low-temperature heat source from electrolysis and methanation (ca. 60–75 °C), which is usually challenging to use with high efficiency [47], could be used to an extent of only 50% in the summer (when P2M operates focusing on energy storage) based on the infrastructure and the expert interviews regarding WWTPs.

3.2.2. Commercial Challenges

Based on the financial analysis results, it can be seen that a 1 MW_{el} P2M plant could operate with minor profitability with an operation time of 1200 h/year at a WWTP, even if it did not pay for the electricity (or it were compensated), and only for system usage. For example, this means only ca. 73,000 EUR profit/year at a biomethane price of 150 EUR/MWh, which is the highest in Europe according to Koonaphapdeelert et al. [37]. Consequently, as the interviews outlined, this business model was not attractive enough for WWTP executives, if they would have to finance the investment costs. According to them, a 7–15 year-long payback period would be favorable. However, even if it were possible, the specific financing questions outlined that WWTPs do not have the financial resources to realize such an investment. For example, the 4,806,000 EUR CAPEX is rather high for a WWTP, if its annual revenue is around 20,000,000 EUR (illustrative data). Moreover, some large rural WWTPs operated unprofitably in previous years, some operated with almost zero balance, and even the profitable ones, which could generate over 500,000 EUR per year, argued that this profit must be handled as retained earnings for unexpected maintenance tasks, not for R&D&I investments.

Even though increasing the number of operating hours could enhance profitability at first glance, other problems would arise:

(1) If a P2M plant—as van Leeuwen and Zauner suggested [45]—were to source electricity from the day-ahead market without any discounts or compensation, one could see that the growing electricity prices in Hungary in recent years do not enhance profitability (the Hungarian Power Exchange Day-Ahead Market Base Average Price was 40 EUR/MWh in 2015 and 5036 EUR/MWh in 2019) [48].

(2) There is some uncertainty as to whether the "bio" prefix, and therefore the premium price, is applicable in the market (outside a national mandatory system) for the output methane gas if only one

input factor comes from renewable sources. There is no consensus in the literature, or in the industry, regarding this question. For example, biomethane is often described as a biogas, the CO_2 content of which is mostly eliminated or separated [49], while green gases are also characterized as a renewable gas [50] that is virtually carbon-neutral [51], made from biomass or with P2G technology [52], but only preferably (and not always) from renewable electricity sources [50]. In one STORE&GO study, "a green gas is defined as a gaseous energy carrier offered to the market without a serious GHG footprint" ([53], p. 12). Even though Jempa et al. [53] pointed out that not only renewable but the nuclear energy can be considered to be carbon-neutral, the concrete business opportunities of such a product gas remain uncertain, mainly because of the currently underdeveloped certificate markets [54] and the missing harmonized and detailed rules on guarantees of origin at the EU level [55].

In sum, neither the characteristics of the seasonal energy storage-focused business model, nor their financial opportunities allow WWTPs to commit to P2M deployment.

3.2.3. Scenarios to Incite WWTPs to Participate in Seasonal Energy Storage

Based on the above, the authors generated scenarios with specific variables including not only electricity sourcing and biomethane price, but public funding for the investment. The goal was to identify the conditions under which the P2M investment could be considered attractive (7–15-year-long payback period) for seasonal energy storage at WWTPs. Specific variables are presented in Table 2. The variables for electricity sourcing are based on the formerly introduced assumption that Transmission System Operators (TSOs) will be forced to avoid energy loss and network imbalance with a framework in which P2M plants can use surplus energy at a favorable or compensated price. The lowest biomethane FiT was generated as a more or less competitive price compared to natural gas, while the highest was based on the highest European FiT (Italy) [37]. The percentage of the public funding of CAPEX was adjusted to the established institutional routines at similar development projects.

Table 2. Specific variables for financially attractive scenario generation for WWTPs.

Financial Factors	Variable 1	Variable 2	Variable 3
Electricity sourcing costs (ESC)	Partly disregarded or compensated: P2M plants do not have to pay for the energy or it is compensated with flexibility/energy storage fees but has to pay the grid power fees for system usage.	Fully disregarded or compensated: P2M plants do not have to pay for the energy, nor for system usage or these are compensated with flexibility/energy storage fees.	
Biomethane FiT (EUR/MWh)	50	100	150
CAPEX support (% of public funding)	50	70	90

Configurations in which the payback period was 7 years and 15 years were explored based on the variable biomethane price. Figure 3 shows that a 7-year-long payback period with 1200 operating hours could be achievable with a reasonable biomethane price (based on international benchmarks [37]) if there were 90% public funding, and even electricity sourcing costs (ESC) were not only partly but fully (including system usage fees) disregarded or compensated (e.g., there was a fee for providing flexibility services or energy storage).

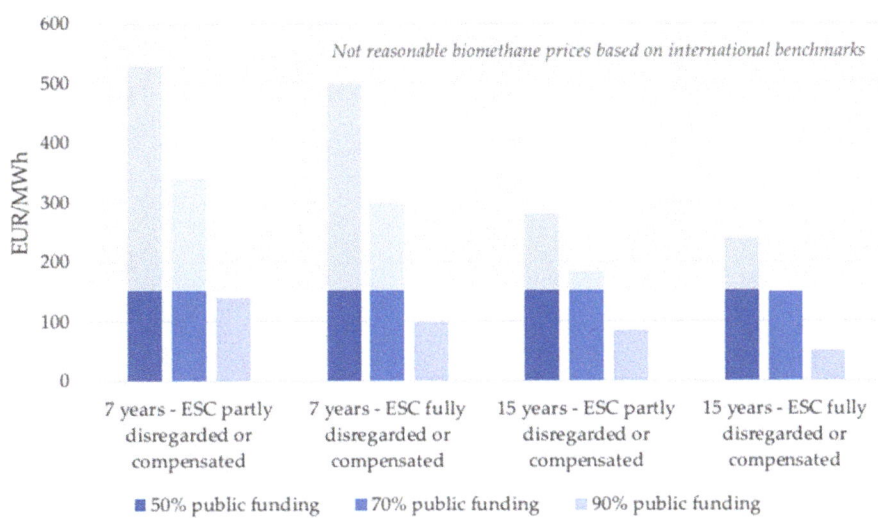

Figure 3. Biomethane prices for 7- and 15-year-long payback periods of a 1 MW$_{el}$ P2M plant focusing only on seasonal energy storage with 1200 operating h/year.

Figure 4 shows that almost 100% public funding is needed to meet the 7-year-long criterion if there is a low biomethane price of 50 EUR/MWh. The lowest public funding percentage is 69% at a biomethane price of 150 EUR/MWh, with fully disregarded or compensated ESC, resulting in a 15-year-long payback period.

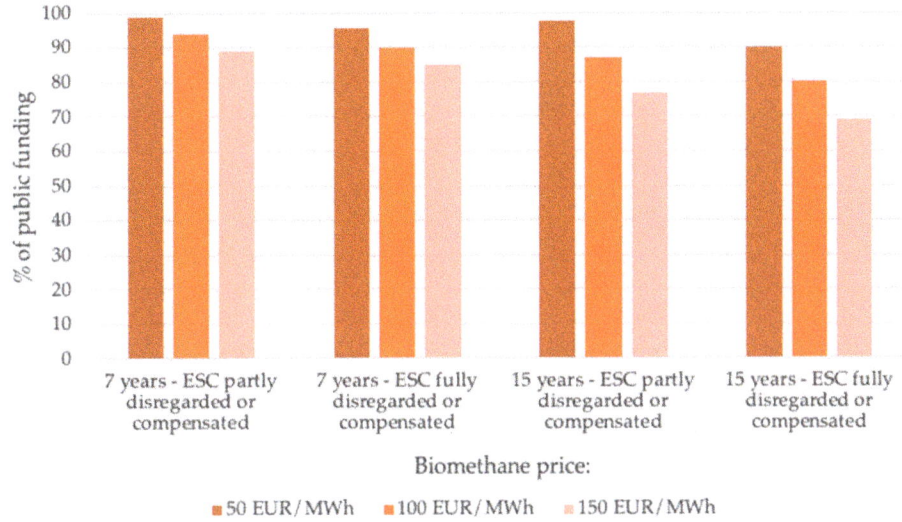

Figure 4. Percentages for public funding of CAPEX for 7- and 15-year-long payback period of a 1 MW$_{el}$ P2M plant focusing only on seasonal energy storage with 1200 operating h/year.

As the executive interviews outlined that Hungarian WWTPs do not have financial resources for a P2M investment, and core activities of WWTPs require stability, prudent risk-management, and efficient operation, they cannot take the innovation-related up-scaling risks and the uncertainties of the business model (currently as potential "first movers" in Hungary) under the current market and regulatory environment. Consequently, public funding is needed to incite WWTPs towards P2M deployment.

For seasonal energy storage with biomethane production, which could be considered an economically beneficial activity in a country that imports ca. 80% of its natural gas [21], dominant public funding for P2M deployment could be justified. Based on the calculations above, the public funding of the CAPEX for 20 × 1 MW$_{el}$ P2M plant would require 66,000,000–93,000,000 EUR, depending on the biomethane FiT, and a framework is also needed within the costs of surplus electricity consumption are minimized or compensated on the revenue side.

4. Discussion

This study focused on techno-economic assessment of the P2M technology deployment in Hungary with a complementary business viewpoint to highlight the concrete opportunities and limitations of seasonal energy storage at WWTPs with biological methanation. The research aimed to answer the following two questions:

(1) What is the total seasonal energy storage potential with P2M at large WWTPs in Hungary?

(2) What are the economic conditions under which WWTPs are financially incited to start the grid-scale implementation of the biological methanation?

Regarding the first research question, the empirical research pointed out that the practical potential of a P2M plant (1 MW$_{el}$) is half of the theoretical potential because of higher methane content and smaller gas flow than expected based on official data and previous research. The 1 MW$_{el}$ P2M size, however, meets the current state of the technology, demonstrated by Electrochaea in Avedøre, Denmark, where the largest P2G plant with biological methanation has been built. As there are around 20 relevant WWTPs exceeding 100,000 PE with biogas production, the total P2M potential at them is around 20 MW$_{el}$, meaning 11.7 GWh seasonal energy storage potential on national level. It could be argued that this volume could be considered part of the decentralized seasonal energy storage system of the country, as the research was focusing on the WWTPs of larger rural cities of Hungary. Considering this potential in a broader context, the national energy strategy plans to reduce the overall natural gas consumption to 8,700,000,000 m^3 (ca. 2550 GWh) to 2030 [21]. With 20 MW$_{el}$ P2M deployments for seasonal energy storage at WWTPs, the 11.7 GWh stored energy could mean ca. 0.5% of the reduced natural gas consumption and equal to the annual energy consumption of ca. 5400 households currently (as the average consumption was 2168 kWh/year/household in 2019 [56]) Though it is not much, at first sight, savings on natural gas import and additional positive externalities (higher integration of renewables, carbon reuse, sector coupling, prevented electricity network imbalances, and related maintenance costs) must be also taken into account. Further research could extend the scope of the financial analyses for these externalities as well.

Regarding the second research question, this energy storage potential can be realized if WWTPs are incited by public funding for P2M deployment and operation, because the current market and regulatory conditions do not meet the criteria of WWTPs for the payback period, and the WWTPs do not have financial resources either to realize a P2M deployment, or to take risks with the still uncertain grid-scale operation and business environment of P2M. This operational uncertainty is derived mainly from the skepticism of WWTP executives, as they have not seen such a plant operating anywhere before, especially not in Hungary. As the National Energy Strategy 2030 plans to support a pilot P2G plant within a few years [21], hopefully, this problem will be solved.

Based on the financial calculations of an "average" WWTP case, the planned mandatory national purchasing system for biomethane by the national energy strategy and the public funding of CAPEX seem not to be enough to incite WWTPs to participate in seasonal energy storage. In other words, while P2M energy storage fits the technological infrastructure of WWTPs, it does not meet their business opportunities and requirements. Currently high and growing electricity prices, through which further PV capacity investment is incited, fundamentally limits the viability of the P2M business model if there are no discounts on the cost side for the consumption of surplus energy or new revenue streams (e.g., aFRR supporting P2G) through which electricity sourcing costs can be compensated.

Nevertheless, one could argue that growing PV capacities (as supply) will suppress electricity market prices, as negative electricity prices have been seen in other European countries (e.g., in Germany) [57]. This could be true in a perfect (in practice: never existing) market and in the long-term, but Hungary's electricity generation from PVs is still low (for example, the annual volume of electricity produced from solar photovoltaic was only 0.02% compared to Germany's production in 2019 [58]). In the short term, with former state intervention to incite PV investment by a FiT system, there is a need for intervention regarding energy storage as well.

Obviously, there is a trade-off between support mechanisms, for example, a larger percentage of public funding of CAPEX can be combined with a lower FiT for biomethane. Based on the generated scenarios and missing financial resources of the WWTPs, there is a clear need for public funding of over 90%. Considering a reasonable FiT for biomethane, other European prices could be referred to in order to contextualize this question: 1.03 EUR/Nm^3 in The Netherlands, 129.7 EUR/MWh in France, 70 EUR/MWh in the UK, 150 EUR/MWh in Italy [37]. Based on these prices, around 75 million EUR CAPEX support and 100 EUR/MWh FiT seem to be the preconditions for realizing the energy storage potential at WWTPs if surplus electricity sourcing costs were also minimized or compensated within a new framework.

The presented results show a significant contribution to the latest literature, as well. For example, while Guerra et al. [59] filled the research gap of the overlooked potential grid benefits of seasonal storage (the literature mainly focused on costs, previously) with their new model for pumped hydro, compressed air, and hydrogen seasonal storage and showed that "for more than 2 days of discharge duration, the only cost-effective technology is hydrogen" [p. 23], this paper emphasized the promising role of methane-based seasonal energy storage if a connection to the natural gas grid is given. Moreover, this research extended the scope of the analysis even more by integrating the motivation and strategic interests of future seasonal energy storage operators and building the financial model on the empirical data of individual sites. Other findings of this paper are in line with the conclusions of latest studies related to global carbon mitigation initiatives [60]. To mitigate environmental damage, Doğan et al. [61] suggest that "OECD governments should directly invest in technological innovation to enhance sustainable economic growth" [p. 9] and Shahzad et al. [62] conclude that "the policymakers of the United States should adopt policies to encourage investors to invest in cleaner energy infrastructure and advanced technologies" [p. 12]. These statements are in line with the conclusion that public funding for P2M seasonal energy storage is essential not only because of the missing capital of WWTPs, but for decreasing GHG emissions, as well. According to Doğan et al., these technological innovations include, however, much more than energy technologies: artificial intelligence and ICT developments could also be mentioned here. These technologies could indicate further possible development projects that would affect the overall WWTP efficiency, and thus, the P2M CAPEX or OPEX in the long term. For example, industrial big data analytics and machine learning [63], which could forecast weather conditions for renewable energy generation (and storage) [64], could become a key success factor (or following Osterwalder and Pigneur's terminology [65], a key resource) in the business model for cost-efficient operation. Furthermore, combining this with the trend towards smart energy systems [32] and technology-driven shared economy [66] could subsequently redefine the role of WWTPs within the rising smart energy communities [67]. These future directions could generate further R&D&I projects which could be also valid for public funding.

5. Conclusions

The hypothesis of this study was the following:

Economic, commercial and investment aspects of P2M seasonal energy storage do not motivate WWTPs to act as future P2M operators, consequently, there is a need for change in the regulatory environment to incite them to realize their seasonal energy storage potential with P2M deployment.

The hypothesis can be accepted, as the results showed that the main criterion of WWTPs for P2M technology investment was the 7–15-year-long payback period. This cannot be achieved in the current market and regulatory environment; possible regulatory changes could affect, however, some of their key motivating factors. To address WWTP stakeholders' expectations, a total of ca. 75 million EUR public funding of CAPEX and 100 EUR/MWh biomethane feed-in tariff is needed to realize their energy storage potential in Hungary if surplus electricity sourcing costs are also minimized or compensated under a new national regulatory framework. The research hypothesis indirectly also suggested that technical aspects would not be hampering factors of P2M technology deployment at large Hungarian WWTPs, which was also proven in this study. The findings show that a standardized 1 MW$_{el}$ P2M technology would fit with most potential sites, and this is in line with the current technology readiness level of P2M.

This study opened new perspectives on techno-economic assessments of P2M technology by integrating not only techno-economic, but also complementary commercial/investment attractiveness of seasonal energy storage at large WWTPs, as well. Due to this approach, the authors could reveal three lessons using Hungary as a case study. First, regarding other economies at similar levels, it is important to highlight that former state interventions inciting new renewable energy generation investments induce a need for intervention on the energy storage side, as well, to avoid loss of surplus energy generation and network imbalance. Second, the research highlighted the 7–15-year-long payback period expectations of future P2M technology operators. Without fulfilling their commercial and investment motivations, any seasonal energy storage initiative will fail. Third, it was shown for the first time (by concrete numbers and proportions) that a three-element regulatory configuration (public funding, FiT, ESC) could have an impact on the attractiveness of P2M seasonal energy storage for WWTPs.

Even though WWTPs could be key for sector coupling and seasonal energy storage, this is only one possible segment of P2M deployment. For example, agricultural biogas plants are also promising because of their on-site CO_2, where the impacts of recent advances in nutrient management to accelerate biogas production [68] could be researched with the P2M process, as well. Further development of carbon capture technologies will bring more flexibility for locating P2M plants. Furthermore, even the lack of a nearby natural gas grid could be bypassed with liquid methane (LNG) and re-gasification [69]. Consequently, examining other or all of the possible P2M deployment segments could be the scope of further research to support policymakers with a more comprehensive analysis.

There are possibilities for further research regarding the method of economic analysis, as well. For example, with respect to a single WWTP or another future P2M technology operator, a complex valuation of the business is needed, by analyzing the business opportunity of the public-funded R&D&I project phase with a limited lifespan [70] and the phase after the mandatory maintenance period of the project with operations on own financial risks. Besides the new tangible assets and perhaps a more favorable market environment, evaluating the acquired intangible assets during an R&D&I project (which could generate premium revenues [71]) could also be a determining factor on whether a WWTP would integrate P2M and seasonal energy storage into their core activities. In line with Machová and Vochozka [72], artificial neural networks could be used not only for the analysis of business companies, but business opportunities to handle these technical, market, and asset valuation complexities of the P2M business case, as well. If site-specific technological complexities would arise because of parallel development projects (e.g., P2M deployment, a capacity increase of a biogas plant, new infrastructure to use oxygen by-product), simulation software like ASPEN PLUS [73] could be applied.

As a concluding remark, the authors hope that their WWTP-focused, in-depth analysis was able to illustrate that there are important commercial and investment viewpoints of future P2M technology operators which should be taken into account to make a step forward with seasonal energy storage towards a more carbon-neutral energy sector.

Author Contributions: Conceptualization, Z.C.; Data collection: Z.C., B.S.-S., M.Z.; Technical analyses: B.S.-S.; Economic analyses: M.Z.; writing: Z.C., B.S.-S., M.Z. All authors have read and agreed to the published version of the manuscript.

Funding: This research received no external funding.

Acknowledgments: The authors would like to thank Hiventures Zrt./State Fund for Research and Development and Innovation for their investment that enabled this research.

Conflicts of Interest: The authors perform their P2G research at Corvinus University of Budapest, and they have founded the innovative startup company Power-to-Gas Hungary Kft. in order to perform industrial R&D and further develop the technology in pre-commercial and commercial environments.

Appendix A

Table A1. Base case for CAPEX calculation at a single WWTP.

Category	Item	Thousand EUR	Unit	Source
Components, physical infrastructure	Electrolyzer system (PEM)	1.6	/kW_{el}	STORE&GO: D8.3. p. 14, 25, 34, 35 D7.5. p. 48
	Methanation system (biological)	0.5	/kW_{el}	
	Infrastructure, installation, storage for gas puffer (H_2, CO_2), injection	1.1	/kW_{el}	
Other	Project development, planning, expert services, quality management	+28%	on costs of total components	Own estimation based on interviews
	Tender-specific R&D, software and maintenance tasks	+50%		

Appendix B

Table A2. Base case for operative expenses and revenues at a single WWTP.

Category	Item	EUR	Unit	Source
Input materials-unit prices	Electricity price	None	-	Disregard based on the fundamental assumption of the study
	Water	0.6	/kW_{el}	Hungarian waterworks
	Power grid fees/ System usage	Variables: None or 1,1	/kW_{el}	Based on Hungarian Energy and Public Utility Regulatory Authority [74]
Operation and maintenance costs	Electrolysis system	4.0%	% of CAPEX at 8000 operating hours	Own estimation based on STORE&GO D8.3. p. 35
	Methanation system	5.0%		
	Infrastructure, installation, storage for gas puffer (H_2, CO_2), injection	3.5%		
Revenues	Biomethane	Variables: 50–150	/MWh	Based on Koonaphapdeelert, et al. [37]
	Waste heat	55	/MWh	STORE&GO D7.7 p. 65
	CO_2 quota	25	/tons	[75]
	Oxygen	0.07	/Nm^3	STORE&GO D7.7 p. 65
Operation data	Operating hours	1200	/year	-
	Directly connected PV capacity	0%	-	Based on WWTP interviews
	Sold/injected biomethane	100%	/total produced	
	Used or sold waste-heat	50%	/total produced	
	Used or sold oxygen	50%	/total produced	

Abbreviations

AD	Anaerobic digestion
AFFR	Automatic frequency restoration reserve
CAPEX	Capital expenditures
CGEN	Combined Gas and Electricity Networks
CHP unit	Combined heat and power unit
EMG-BES	Bioelectrochemical system for electromethanogenesis
ESC	Electricity sourcing costs
FiT	Feed-in tariff
GHG	Greenhouse gas
HHV	Higher heating value
LNG	Liquefied natural gas
OPEX	Operating expenditures
P2G	Power-to-gas
P2H	Power-to-hydrogen
PM	Power-to-methane
PE	Population Equivalent
PEMEC	Polymer electrolyte membranes electrolysis
PV	Photovoltaics
R&D&I	Research, development, and innovation
SNG	Synthetic Natural Gas
TSO	Transmission System Operator
WWTP	Wastewater treatment plant

References

1. Bailera, M.; Lisbona, P.; Romeo, L.; Espatolero, S. Power to Gas Projects Review: Lab, Pilot and Demo Plants for Storing Renewable Energy and CO_2. *Renew. Sustain. Energy Rev.* **2017**, *69*, 292–312. [CrossRef]
2. Blanco, H.; Faaij, A. A Review at The Role of Storage in Energy Systems with A Focus on Power to Gas and Long-Term Storage. *Renew. Sustain. Energy Rev.* **2018**, *81*, 1049–1086. [CrossRef]
3. Csedő, Z.; Zavarkó, M. The role of inter-organizational innovation networks as change drivers in commercialization of disruptive technologies: The case of power-to-gas. *Int. J. Sustain. Energy Plan. Manag.* **2020**, *28*, 53–70. [CrossRef]
4. Jäger-Waldau, A. *PV Status Report 2019*; Publications Office of the European Union: Luxembourg, 2019. [CrossRef]
5. Sarkar, D.; Odyuo, Y. An ab initio issues on renewable energy system integration to grid. *Int. J. Sustain. Energy Plan. Manag.* **2019**, *23*, 27–38. [CrossRef]
6. Schiebahn, S.; Grube, T.; Robinius, M.; Tietze, V.; Kumar, B.; Stolten, D. Power to Gas: Technological Overview, Systems Analysis and Economic Assessment for A Case Study in Germany. *Int. J. Hydrogen Energy* **2015**, *40*, 4285–4294. [CrossRef]
7. Götz, M.; Lefebvre, J.; Mörs, F.; McDaniel Koch, A.; Graf, F.; Bajohr, S.; Reimert, R.; Kolb, T. Renewable Power-To-Gas: A Technological and Economic Review. *Renew. Energy* **2016**, *85*, 1371–1390. [CrossRef]
8. Schäfer, M.; Gretzschel, O.; Steinmetz, H. The Possible Roles of Wastewater Treatment Plants in Sector Coupling. *Energies* **2020**, *13*, 2088. [CrossRef]
9. Gretzschel, O.; Schäfer, M.; Steinmetz, H.; Pick, E.; Kanitz, K.; Krieger, S. Advanced Wastewater Treatment to Eliminate Organic Micropollutants In Wastewater Treatment Plants in Combination With Energy-Efficient Electrolysis At WWTP Mainz. *Energies* **2020**, *13*, 3599. [CrossRef]
10. Ceballos-Escalera, A.; Molognoni, D.; Bosch-Jimenez, P.; Shahparasti, M.; Bouchakour, S.; Luna, A.; Guisasola, A.; Borràs, E.; Della Pirriera, M. Bioelectrochemical Systems For Energy Storage: A Scaled-Up Power-To-Gas Approach. *Appl. Energy* **2020**, *260*, 114138. [CrossRef]
11. Maroušek, J.; Stehel, V.; Vochozka, M.; Kolář, L.; Maroušková, A.; Strunecký, O.; Peterka, J.; Kopecký, M.; Shreedhar, S. Ferrous Sludge from Water Clarification: Changes in Waste Management Practices Advisable. *J. Clean. Prod.* **2019**, *218*, 459–464. [CrossRef]

12. Maroušek, J.; Kolář, L.; Strunecký, O.; Kopecký, M.; Bartoš, P.; Maroušková, A.; Cudlínová, E.; Konvalina, P.; Šoch, M.; Moudrý, J., Jr.; et al. Modified Biochars Present An Economic Challenge To Phosphate Management In Wastewater Treatment Plants. *J. Clean. Prod.* **2020**, *272*, 123015. [CrossRef]
13. Ghaib, K.; Ben-Fares, F. Power-To-Methane: A State-Of-The-Art Review. *Renew. Sustain. Energy Rev.* **2018**, *81*, 433–446. [CrossRef]
14. Ameli, H.; Qadrdan, M.; Strbac, G. Techno-Economic Assessment of Battery Storage and Power-To-Gas: A Whole- System Approach. *Energy Procedia* **2017**, *142*, 841–848. [CrossRef]
15. Collet, P.; Flottes, E.; Favre, A.; Raynal, L.; Pierre, H.; Capela, S.; Peregrina, C. Techno-Economic and Life Cycle Assessment of Methane Production Via Biogas Upgrading and Power to Gas Technology. *Appl. Energy* **2017**, *192*, 282–295. [CrossRef]
16. Peters, R.; Baltruweit, M.; Grube, T.; Samsun, R.; Stolten, D. A Techno Economic Analysis of The Power to Gas Route. *J. Co2 Util.* **2019**, *34*, 616–634. [CrossRef]
17. Maroušek, J.; Strunecký, O.; Stehel, V. Biochar Farming: Defining Economically Perspective Applications. *Clean Technol. Environ. Policy* **2019**, *21*, 1389–1395. [CrossRef]
18. Maroušek, J.; Rowland, Z.; Valášková, K.; Král, P. Techno-Economic Assessment of Potato Waste Management in Developing Economies. *Clean Technol. Environ. Policy* **2020**, *22*, 937–944. [CrossRef]
19. Mardoyan, A.; Braun, P. Analysis of Czech Subsidies for Solid Biofuels. *Int. J. Green Energy* **2014**, *12*, 405–408. [CrossRef]
20. Eveloy, V.; Gebreegziabher, T. A Review of Projected Power-To-Gas Deployment Scenarios. *Energies* **2018**, *11*, 1824. [CrossRef]
21. Ministry for Innovation and Technology of Hungary. *National Energy Strategy 2030*; Ministry for Innovation and Technology of Hungary: Budapest, Hungary, 2020.
22. Mazza, A.; Bompard, E.; Chicco, G. Applications of Power to Gas Technologies in Emerging Electrical Systems. *Renew. Sustain. Energy Rev.* **2018**, *92*, 794–806. [CrossRef]
23. Fasihi, M.; Bogdanov, D.; Breyer, C. Techno-Economic Assessment of Power-to-Liquids (PtL) Fuels Production and Global Trading Based on Hybrid PV-Wind Power Plants. *Energy Procedia* **2016**, *99*, 243–268. [CrossRef]
24. Ferry, J.G. Enzymology of One-Carbon Metabolism in Methanogenic Pathways. *Fems Microbiol. Rev.* **1999**, *23*, 13–38. [CrossRef] [PubMed]
25. Mets, L. Methanobacter Thermoautotrophicus Strain and Variants. Thereof. Patent EP2661511B1, 5 January 2012.
26. Lovato, G.; Alvarado-Morales, M.; Kovalovszki, A.; Peprah, M.; Kougias, P.G.; Rodrigues, J.A.D.; Angelidaki, I. In-Situ Biogas Upgrading Process: Modeling and Simulations Aspects. *Bioresour. Technol.* **2017**, *245*, 332–341. [CrossRef] [PubMed]
27. Ministry of Interior of Hungary. *Information on the National Implementation Program of Directive 91/271/EEC About Waste Water Drainage and Treatment of Settlements of Hungary*; Ministry of Interior of Hungary: Budapest, Hungary, 2018.
28. Adil, A.M.; Ko, Y. Socio-Technical Evolution of Decentralized Energy Systems: A Critical Review and Implications for Urban Planning and Policy. *Renew. Sustain. Energy Rev.* **2016**, *57*, 1025–1037. [CrossRef]
29. Act, No. XL of 2008 on Natural Gas Supply, Hungary. Available online: http://njt.hu/cgi_bin/njt_doc.cgi?docid=117673.386422 (accessed on 14 August 2020).
30. Governmental Decree No. 19 of 2009 (I.30.) Implementing the Provisions of the Act, Hungary. Available online: http://njt.hu/cgi_bin/njt_doc.cgi?docid=124061.388348/ (accessed on 14 August 2020).
31. STORE&GO. The Project STORE&GO—Shaping the Energy Supply for the Future. Available online: https://www.storeandgo.info/about-the-project/ (accessed on 21 February 2020).
32. Lund, H.; Østergaard, P.A.; Connolly, D.; Ridjan, I.; Mathiesen, B.V.; Hvelplund, F.; Thellufsen, J.Z.; Sorknæs, P. Energy Storage and Smart Energy Systems. *Int. J. Sustain. Energy Plan. Manag.* **2016**, *11*, 3–14. [CrossRef]
33. Pintér, G.; Zsiborács, H.; Baranyai, N.H.; Vincze, A.; Birkner, Z. The Economic and Geographical Aspects of the Status of Small-Scale Photovoltaic Systems in Hungary—A Case Study. *Energies* **2020**, *13*, 3489. [CrossRef]
34. Thema, M.; Weidlich, T.; Hörl, M.; Bellack, A.; Mörs, F.; Hackl, F.; Kohlmayer, M.; Gleich, J.; Stabenau, C.; Trabold, T.; et al. Biological CO_2-Methanation: An Approach To Standardization. *Energies* **2019**, *12*, 1670. [CrossRef]
35. Varone, A.; Ferrari, M. Power to Liquid and Power to Gas: An Option for the German Energiewende. *Renew. Sustain. Energy Rev.* **2015**, *45*, 207–218. [CrossRef]

36. Solargis. Solar Resource Maps of Hungary. Available online: https://solargis.com/maps-and-gis-data/download/hungary (accessed on 15 August 2020).
37. Koonaphapdeelert, S.; Aggarangsi, P.; Moran, J. *Biomethane: Production and Applications*; Springer: Singapore, 2020.
38. Glaser, B.G.; Strauss, A.L. *The Discovery of Grounded Theory: Strategies for Qualitative Research*; Aldine de Gruyter: Chicago, IL, USA, 1967.
39. Corbin, J.M.; Strauss, A.L. *Basics of Qualitative Research: Techniques and Procedures for Developing Grounded Theory*; SAGE: Los Angeles, CA, USA, 2008.
40. Zhang, Y.; Wildemuth, B.M. Qualitative analysis of content. *Hum. Brain Mapp.* **2005**, *30*, 2197–2206.
41. Bai, A. *A Biogáz (The Biogas)*; Száz Magyar Falu Könyvesháza: Budapest, Hungary, 2007.
42. Kisari, K. Economic Assessment and Organizational Aspects of the Applicability of the LEAN Model in Biogas Plants. Ph.D. Thesis, Szent István University, Gödöllő, Hungary, 2017.
43. Sinóros-Szabó, B. Evaluation of Biogenic Carbon Dioxide Market and Synergy Potential for Commercial-Scale Power-to-Gas Facilities in Hungary. In *Proceedings of Innovációs Kihívások a XXI. Században (Innovation Challenges in the XXI. Century), LXI. Georgikon Days Conference, 3–4 October 2019*; Pintér, G., Csányi, S., Zsiborács, H., Eds.; University of Pannonia–Georgikon Faculty: Keszthely, Hungary, 2019; pp. 371–380. ISBN 9789633961308.
44. Böhm, H.; Zauner, A.; Goers, S.; Tichler, R.; Pieter, K. *D7.5. Report on Experience Curves and Economies of Scale*; STORE&GO: Karlsruhe, Germany, 2018.
45. Van Leeuwen, C.; Zauner, A.D. *8.3. Report on the Costs Involved with PtG Technologies and Their Potentials Across the EU*; STORE&GO: Karlsruhe, Germany, 2018.
46. Zauner, A.; Böhm, H.; Rosenfeld, D.C.; Tichler, R. *D7.7. Analysis on Future Technology Options and on Techno-Economic Optimization*; STORE&GO: Karlsruhe, Germany, 2019.
47. Györke, G.; Groniewsky, A.; Imre, A. A Simple Method of Finding New Dry and Isentropic Working Fluids for Organic Rankine Cycle. *Energies* **2019**, *12*, 480. [CrossRef]
48. Hungarian Power Exchange. Annual and Monthly Reports. Available online: https://hupx.hu/hu/piaci-adatok/dam/rendszeres-riportok (accessed on 18 August 2020).
49. European Biogas Association. About Biogas and Biomethane. Available online: https://www.europeanbiogas.eu/about-biogas-and-biomethane/ (accessed on 18 June 2020).
50. Murphy, J.D. *Green Gas Facilitating a Future Green Gas Grid through the Production of Renewable Gas*; (eBook electronic edition); IEA Bioenergy: Paris, France, 2018; ISBN 978-1-910154-38-0.
51. Ecotricity. What is Green Gas? Available online: https://www.ecotricity.co.uk/our-green-energy/our-green-gas/what-is-green-gas (accessed on 18 June 2020).
52. Junginger, M.; Baxter, D. *Biomethane—Status and Factors Affecting Market Development and Trade*, electronic version; IEA Bioenergy: Paris, France, 2014; ISBN 978-1-910154-10-6.
53. Jepma, C.; van Leeuwen, C.; Hulshof, D. *D8.1. Exploring the Future for Green Gases*; STORE&GO: Karlsruhe, Germany, 2017.
54. Hulshof, D.; Jempa, C.; Mulder, M. *D8.2. Report on the Acceptance and Future Acceptability of Certificate-Based Green Gases*; STORE&GO: Karlsruhe, Germany, 2018.
55. Kreeft, G.J. *D7.2. European Legislative and Regulatory Framework on Power-to-Gas*; STORE&GO: Karlsruhe, Germany, 2017.
56. NKM National Utilities. Átlagos Éves Fogyasztás (Average Annual Consumption). Available online: https://www.nkmenergia.hu/aram/pages/aloldal.jsp?id=550565 (accessed on 15 September 2020).
57. Vos, K.D. Negative Wholesale Electricity Prices in the German, French and Belgian Day-Ahead, Intra-Day and Real-Time Markets. *Electr. J.* **2015**, *28*, 36–50. [CrossRef]
58. Statista. Annual Volume of Electricity Produced from Solar Photovoltaic in the European Union (EU-28) in 2019, by Country. Available online: https://www.statista.com/statistics/863238/solar-photovoltaic-power-electricity-production-volume-european-union-eu-28/ (accessed on 15 August 2020).
59. Guerra, O.; Zhang, J.; Eichman, J.; Denholm, P.; Kurtz, J.; Hodge, B. The Value of Seasonal Energy Storage Technologies for The Integration of Wind and Solar Power. *Energy Environ. Sci.* **2020**, *13*, 1909–1922. [CrossRef]

60. Sarwar, S.; Shahzad, U.; Chang, D.; Tang, B. Economic and Non-Economic Sector Reforms in Carbon Mitigation: Empirical Evidence from Chinese Provinces. *Struct. Chang. Econ. Dyn.* **2019**, *49*, 146–154. [CrossRef]
61. Doğan, B.; Driha, O.M.; Balsalobre Lorente, D.; Shahzad, L. The mitigating effects of economic complexity and renewable energy on carbon emissions in developed countries. *Sustain. Dev.* **2020**, *1*, 1–12. [CrossRef]
62. Shahzad, U.; Fareed, Z.; Shahzad, F.; Shahzad, K. Investigating the Nexus Between Economic Complexity, Energy Consumption and Ecological Footprint for The United States: New Insights from Quantile Methods. *J. Clean. Prod.* **2020**, *279*, 123806. [CrossRef]
63. Graessley, S.; Šuleř, P.; Klieštik, T.; Kicová, E. Industrial Big Data Analytics for Cognitive Internet of Things: Wireless Sensor Networks, Smart Computing Algorithms, And Machine Learning Techniques. *Anal. Metaphys.* **2019**, *18*, 23–29. [CrossRef]
64. Scher, S.; Messori, G. Predicting Weather Forecast Uncertainty with Machine Learning. *Q. J. R. Meteorol. Soc.* **2018**, *144*, 2830–2841. [CrossRef]
65. Osterwalder, A.; Pigneur, Y. *Business Modell Generation*; John Wiley & Sons: Hoboken, NJ, USA, 2010.
66. Graessley, S.; Horák, J.; Kováčová, M.; Valášková, K.; Poliak, M. Consumer Attitudes and Behaviors in The Technology-Driven Sharing Economy: Motivations for Participating in Collaborative Consumption. *J. Self Gov. Manag. Econ.* **2019**, *7*, 25–30. [CrossRef]
67. Zhou, S.; Hu, Z.; Gu, W.; Jiang, M.; Zhang, X. Artificial Intelligence Based Smart Energy Community Management: A Reinforcement Learning Approach. *Csee J. Power Energy Syst.* **2019**, *5*, 1–10. [CrossRef]
68. Maroušek, J.; Strunecký, O.; Kolář, L.; Vochozka, M.; Kopecký, M.; Maroušková, A.; Batt, J.; Poliak, M.; Šoch, M.; Bartoš, P.; et al. Advances In Nutrient Management Make It Possible To Accelerate Biogas Production And Thus Improve The Economy Of Food Waste Processing. *Energy Sour. Part A Rec. Util. Environ. Eff.* **2020**, 1–10. [CrossRef]
69. Imre, A.R.; Kustán, R.; Groniewsky, A. Thermodynamic Selection of the Optimal Working Fluid for Organic Rankine Cycles. *Energies* **2019**, *12*, 2028. [CrossRef]
70. Vochozka, M.; Rowland, Z.; Suler, P. The specifics of valuating a business with a limited lifespan. *Ad Alta J. Interdiscip. Res.* **2019**, *9*, 339–345.
71. Stehel, V.; Rowland, Z.; Marecek, J. Valuation of intangible asset deposit into capital company in case of specific transaction. *Ad Alta J. Interdiscip. Res.* **2019**, *9*, 287–291.
72. Machová, V.; Vochozka, M. Analysis of Business Companies Based on Artificial Neural Networks. *Shs Web Conf.* **2019**, *61*, 01013. [CrossRef]
73. Nikoo, M.; Mahinpey, N. Simulation of Biomass Gasification in Fluidized Bed Reactor Using ASPEN PLUS. *Biomass Bioenergy* **2008**, *32*, 1245–1254. [CrossRef]
74. Hungarian Energy and Public Utility Regulatory Authority, "Decree 15/2016. (XII. 20.). 2016. Available online: http://njt.hu/cgi_bin/njt_doc.cgi?docid=199406.376281 (accessed on 14 August 2020).
75. Insider Inc. CO2 European Emission Allowances. Available online: https://markets.businessinsider.com/commodities/co2-european-emission-allowances (accessed on 14 August 2020).

© 2020 by the authors. Licensee MDPI, Basel, Switzerland. This article is an open access article distributed under the terms and conditions of the Creative Commons Attribution (CC BY) license (http://creativecommons.org/licenses/by/4.0/).

Article

The Potential Role of Power-to-Gas Technology Connected to Photovoltaic Power Plants in the Visegrad Countries—A Case Study

Gábor Pintér

Renewable Energy Research Group, Soós Ernő Research and Development Center, Nagykanizsa Campus, Faculty of Engineering, University of Pannonia, 8800 Nagykanizsa, Hungary; pinter.gabor@uni-pen.hu; Tel.: +36-30-373-8550

Received: 12 November 2020; Accepted: 2 December 2020; Published: 4 December 2020

Abstract: With the spread of the use of renewable sources of energy, weather-dependent solar energy is also coming more and more to the fore. The quantity of generated electric power changes proportionally to the intensity of solar radiation. Thus, a cloudy day, for example, greatly reduces the amount of electricity produced from this energy source. In the countries of the European Union solar power plants are obligated to prepare power generation forecasts broken down to 15- or 60-min intervals. The interest of the regionally responsible transmission system operators is to be provided with forecasts with the least possible deviation from the actual figures. This paper examines the Visegrad countries' intraday photovoltaic forecasts and their deviations from real power generation based on the photovoltaic power capacity monitored by the transmission system operators in each country. The novelty of this study lies in the fact that, in the context of monitored PV capacities in the Visegrad countries, it examines the regulation capacities needed for keeping the forecasts. After comparing the needs for positive and negative regulation, the author made deductions regarding storage possibilities complementing electrochemical regulation, based on the balance. The paper sought answers concerning the technologies required for the balancing of PV power plants in the examined countries. It was established that, as a result of photovoltaic power capacity regulation, among the four Visegrad countries, only the Hungarian transmission system operator has negative required power regulation, which could be utilized in power-to-gas plants. This power could be used to produce approximately 2.1 million Nm^3 biomethane with a 98% methane content, which could be used to improve approximately 4 million Nm^3 biogas of poor quality by enriching it (minimum 60% methane content), so that it can be utilized. The above process could enhance the viability of 4–6 low-methane agricultural biogas plants in Hungary.

Keywords: power-to-gas; regulation; energy storage; biogas; biomethane

1. Introduction

1.1. The Use of Renewable Energy Worldwide

As the world's hunger for energy does not cease to grow, the exploitation of renewable sources of energy is becoming more and more important. For the sustainability of the electric power system (EPS), the efficient integration of variable renewable energy (VRE) is an urgent matter. In the year 2018, VRE-based technologies accounted for 33% of the global electric power generation capacity and more than 26% of total electric power generation [1,2].

The European Union (EU), which is a leading figure in climate policy as well as in the use of renewable energy and the efforts for a new energy economy, has set the target to significantly reduce the emission of greenhouse gases in the next few decades. According to some forecasts, by 2040 about

40–60% of the electric power generated in the EU will come from photovoltaic (PV) and wind power systems. Such a great proportion of VRE requires the electric power system to be flexible enough to cope with weather-dependent energy production. Consequently, it is necessary to possess enough reserve capacity for the periods when the weather conditions do not allow stable VRE generation. It is also predicted that the dynamic spread of these technologies will cause technical problems in the macro-energy systems (e.g., loop flows in the local grid) more and more frequently. For example, this may also mean the optimization of the electric parameters of an entire area served by a transformer station [3–12].

These issues could be solved by the intensive development of energy storage technologies in the European Union. This is especially important, since the establishment of flexible local energy storage capacities has already become a key factor of energy security for the EU member countries due to the spread of VRE [4,11,13–20].

At the beginning of the year 2017, the global capacity of permanent energy storage systems amounted to 159 GW, 153 GW of which belonged to pumped hydro storage (PHS). The rest of the energy storage technologies represented a total of 5.9 GW, of which the share of battery-based devices was 2.3 GW [21–24]. The European PHS capacity reached 50.5 GW in 2017 (approx. 1.9 TWh energy capacity according to [25–29], with more than 59% of this capacity belonging to 5 countries (Italy, France, Germany, Austria, and Spain) [25–27,30,31].

Based on the current trends, it is expected that the European Union's estimated nominal storage power capacity will grow to 72–95 GW, while its energy storage capacity will reach 3.6–4.1 TWh by 2040 [24,26,32,33]. Although, at present, the regulations of energy storage are not unified in the nations of the European Union, the increase in VRE capacities will call for a new regulatory and economic environment for the security of the electricity supply and the spread of energy storage technologies [9,32,34].

As the utilization of variable renewable energy (VRE) had gained more significance in transforming the energy systems of the world, solar energy and solar energy schemes started to play an increasingly prominent role in supporting sustainable development and the protection of the environment. The most apparent reason for turning to solar energy is that—besides the fact that the energy of the Sun is the basis of most processes in nature—it is a readily available, clean, plentiful and sustainable resource [35–44]. To illustrate how abundant this energy source is, it suffices to mention that the potential of the energy from the Sun reaching the Earth annually exceeds humankind's current need for energy several thousand times. Regarding the different ways of making use of this energy, photovoltaic (PV) technology is a common solution, in which PV cells convert the radiation of the Sun into DC energy [45–47]. Nevertheless, it must also be stated here that the diverse solar energy sources that could be available to humanity have not been fully explored and utilized yet [48].

The PV technologies that are most used around the world at present are the amorphous silicon (a-Si), the monocrystalline (m-Si) and the polycrystalline (p-Si) technologies. Probably because of their outstanding reliability, crystalline solar modules are the most common ones, boasting a market share of approximately 90%. The best efficiency that can be reached with p-Si modules is about 26.7%, while m-Si PV ones are capable of approximately 22.3% [49–58]. Nevertheless, the efficiency of the m-Si and p-Si modules that are mostly used today lags behind to a great degree, and remain between 10–18% in the area of the EU [59]. Concerning the a-Si photovoltaic technology, which is a thin-film-based PV technology, the highest value of efficiency to be reached currently is only 10.5%, which is still about twice as high as the 4 to 6% of the a-Si modules mainly used today. As there is no data currently available about the market share of a-Si technology, one can only make assumptions on the basis of the fact that the total share of all thin-film solar modules constitutes about 10% [49,52,53,57–61].

In the past ten years, the PV sector has been witness to an unprecedented growth, which can be explained, on the one hand, by great technological advancements and developments and a plethora of different measures, for example novel financial support schemes, the Feed-in-Tariff and the decreasing expenses related to the investments, on the other hand [39,49,62]. Moreover, hybrid renewable energy microgrids (HREM), one of the central elements of which are PV systems, represent an affordable, reliable and sustainable alternative for providing households with a modern energy supply [63].

Thanks to the dynamic development, three years ago, in 2017, 26.5% of the total amount of electric power produced in the world was generated from renewable sources. The proportion of electricity from PV technologies was 1.9%, which was generated by a worldwide photovoltaic capacity of 402 GW. The first four top PV power producers were China (131.1 GW) in the first place, followed by the European Union (108 GW), the United States of America (51 GW) and Japan (49 GW). In connection with China, it deserves to be noted that as a result of the changes in the roles of the various technologies, PV technology became its most significant new power capacity [39,64].

The amount of energy that can be generated by PV technology depends on many factors, the most important of which is solar radiation followed by the given technology used, the natural conditions of the environment, the temperature, the composition of the particular module and the joint effects of the configuration of the system itself and its efficiency. The map presenting the PV power potentials of Europe clearly shows that the yearly quantity of PV energy that can be generated ranges between 700–1900 kWh/kWp on average, depending on the actual location (Figure 1). Among the four countries of the Visegrad Group, Hungary has the highest figures: 1050–1300 kWh/kWp, followed by Slovakia with 1050–1250 kWh/kWp, while the lowest values belong to Poland and the Czech Republic: 950–1200 kWh/kWp (Figure 1) [58,65–68].

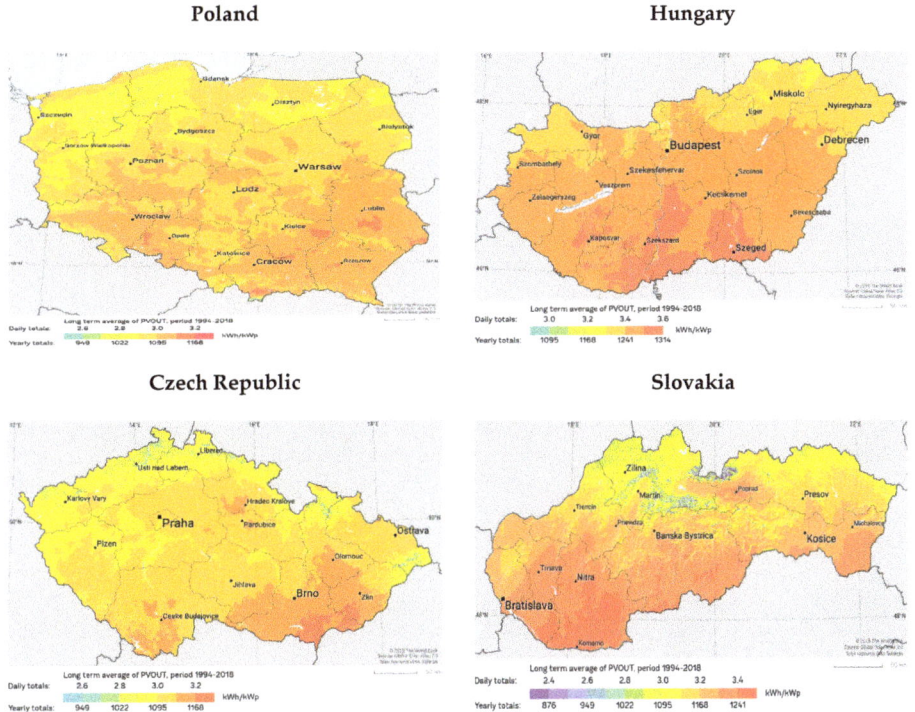

Figure 1. The photovoltaic power potentials of the V4 countries [67].

1.2. Utilizing Biogas and the Potentials of Power-to-Gas Technology

The utilization of biogas dates back to ancient times. It is produced in agriculture as a byproduct, and the last two decades have seen a great rise in its production as well as use in numerous countries, increasing its significance as a renewable source of energy. In the European Union biogas production grew by nearly 100 TWh between 2008 and 2016, when the production reached 187 TWh [69].

Kampman et al. [70] predicted a substantial increase till 2030. The realization of this potential, however, will necessitate joint efforts by a great number of various actors, including those involved in policymaking. They will be the ones who are responsible for creating incentives and initiatives, on the one hand, and eliminating the obstacles in the way of increasing the production and more widespread use of biogas, on the other hand. The institutional conditions for biogas solutions are very complex. Regarding agricultural biogas plants, it has to be mentioned here that they mainly serve an ancillary function rather than being the actual purpose of farming [71]. The wrong composition of raw materials can have an adverse effect on the efficiency of the production of the biogas plant, i.e., the methane content of the gas produced may be too low [72]. Power-to-gas technology may be able to provide a solution to this problem.

In the power-to-gas process, the power-to-hydrogen phase is followed by the power-to-methane stage, which utilizes the hydrogen produced in the first phase with the help of methanation. The biomethane formed by the methanation process can be fed into the natural gas grid in unlimited quantities since its characteristics are the same as those of natural gas.

The two key elements of the power-to-gas process are electrolysis and methanation:

1. Technology uses alkaline, PEME (proton exchange membrane electrolysis) and solid oxide electrolysis methods. Of the three techniques, the alkaline one has been in use for the longest time. The PEME method is more favorable to be used with weather-dependent sources of renewable energy, such as solar energy, because it can support a more flexible system, for example by starting up more quickly. In addition, it is capable of an approximately 5% higher operational efficiency [11,73]. Solid oxide electrolysis requires even less electric power compared to PEME, but its system stability is lower, while its heat requirement is higher [73].
2. Biological and catalytic methanation are typically used in power-to-gas technology. The catalytic (also known as chemical or Sabatier) method has been used since as early as the 1970s. Nevertheless, biological methanation is more preferable because it allows an approximately 20% higher carbon dioxide conversion rate than in the case of the catalytic procedure [74]. The biological method is more flexible (i.e., it can be started more quickly, for example), and its pressure and heat requirements are also lower than those of the catalytic process. For use with VRE sources, biological methanation is recommended due to its greater operational flexibility [75,76].

Both industrial and scientific actors agree that power-to-gas technology has a significant potential for the future [73,74]. Currently, however, there are only a few industrial-size power-to-gas plants around the world, so this technology is still in a phase of initial growth [77,78].

An earlier study of mine was based on five years' data of the Belgian Elia Group, and it monitored the balancing capacities necessary for the regulation of photovoltaic power plants at quarter hourly intervals. The goal of the examination was to establish what the effects of the day-ahead and intraday schedules were on the required regulation [78]. The present study takes a step further from the previous results and only concentrates on the required regulation needed for keeping the intraday schedules in the Visegrad countries, pointing out the possibilities of balancing solar power plants with the help of power-to-gas technology.

2. Methods and Details of the Study

2.1. The Scope of the Investigation

This study deals with the Visegrad countries (Poland, the Czech Republic, the Slovak Republic and Hungary), all of which are members of the European Network of Transmission System Operators for Electricity (ENTSO-E). In order to be as up-to-date as possible, a very recent period, 1 September 2019–31 August 2020, was selected for the purposes of the investigation, which involved the comparison of the real electricity production data with the intraday forecasts of the PV power generation in each Visegrad country. The database used was that of the ENTSO-E, which includes 35 European countries. It was the European Union's Third Legislative Package for the Internal Energy Market that created the ENTSO-E and gave it legal mandates in 2009 with a view to further liberalizing the electricity and gas markets in the Union. As a result of the Third Energy Package, the transmission system operators' roles changed remarkably. Because of the unbundling and liberalization measures in the energy market, TSOs became something like a location in the market for the different players to meet and interact. It is the common goal of the members of the ENTSO-E to establish an internal energy market and guarantee its best possible operation as well as to further the energy and climate goals of the Union. A truly significant challenge the TSOs are faced with currently is the integration of an increased proportion of renewable energy sources in the energy systems of the EU, which involves the enhancement of flexibility as well as a lot more customer-focused approach than ever before.

The first goal in the investigation was to determine the balancing requirements for each country by comparing the positive and negative regulation needs considering the time series. The corresponding positive and negative regulation requirements were analyzed in each forecasting interval. The deviation of real data was calculated relative to the forecasts related to a 100 MWp PV system for the Visegrad countries. The characteristics of the energy consumption of the individual countries were not taken into account; the PV power generation forecasts were compared to the actual PV generation figures.

For the implementation of any PV integration, it is necessary to make country-specific surveys of the amounts of regulation resulting from the deviations from the power plants' forecasts, using the available data. In turn, these data can help with the selection of the suitable energy storage technology and strategy for the electric energy system of a given country.

In the present study, only the intraday forecast data were examined, since it was assumed that these provided more accurate predictions about the expected production compared to the day-ahead forecasts. By mapping the discrepancies between the intraday forecasts and the actual production, the goal was to spot the niches for various energy storage devices, especially power-to-gas ones, to complement the PV capacities of the examined countries for a better compliance with the generation forecasts.

As among the studied countries, only Hungary was found to have negative required power regulation, and since—similarly to Holland—this country also possesses a natural gas infrastructure, which is considered highly developed for European standards [79], it is an obvious solution to utilize the existing network. Conversely, the hydrogen market is still very underdeveloped with hardly any infrastructure. Hungary's National Energy Strategy 2030 [80] supports the same view, as it primarily considers technologies based on the use of the existing natural gas infrastructure besides electrochemical energy storage till 2030.

2.2. The Data Used in the Calculations

The calculations were primarily based on data from the databases of the ENTSO-E and the Hungarian Independent Transmission Operator Company Ltd. (MAVIR ZRT., Budapest, Hungary). The PV power data (most recent forecast and measured data) in the ENTSO-E database are given at intervals of either 15 or 60 min, according to the provision of data in the particular countries [81–83]. Correspondingly, 15-min data were used where available (Hungary and Poland), while in the cases where only hourly data were obtainable, those had to be applied (Slovakia and the Czech Republic) (Table 1). The longer intervals (60-min) give the individual actors in the market more freedom in creating and managing their forecasts than the 15-min ones, since the latter can only be planned on the basis of much more precise meteorological forecasts. Throughout the calculations, the countries with the 15-min and the 60-min intervals were treated separately.

The monitored capacities of the Visegrad countries are shown in Table 2. As it can be seen, the highest figure belongs to the Czech Republic, while the lowest one to Slovakia. For Poland no data were available on 1 September 2019 yet, only starting from May 2020. For the sake of comparability, the data were homogenized, so they were recalculated for 100 MWp PV systems for the individual countries.

Table 1. Intraday photovoltaic (PV) forecast data in the Visegrad countries [83,84].

Country	Availability of PV Forecast	Data Resolution (min)
Czech Republic		60
Hungary	Intraday PV forecast data are available	15
Poland		15
Slovakia		60

Table 2. The monitored PV capacities of the Visegrad countries [83,84].

Country	Size of the Monitored PV Capacity (MWp)	
	1 September 2019	31 August 2020
Czech Republic	2054	2061
Hungary	1013	1129
Poland	NDA	1928
Slovakia	409	450

It is worthy of note that according to Article 5 (*Balance responsibility*) of Regulation (EU) 2019/943 of the European Parliament and of the Council of 5 June 2019 on the internal market for electricity: "All market participants shall be responsible for the imbalances they cause in the system ("balance responsibility")" [85], i.e., the creation of accurate forecasts and consequently the minimization of deviation from it affects every PV power producer in the European Union. Thus, the complete balancing in both a negative and a positive direction needs to be dealt with for every country specifically because of the obligation to comply with PV generation forecasts.

2.3. The Calculations

Obviously, adherence to the PV forecasts is of paramount importance at all times. However, if the actual power generated by a particular PV system during a 15-min or 60-min time period is less than the corresponding value in the intraday forecast, the TSO is obligated to regulate the situation in a positive direction [86]. Thus, from the point of view of the operator of the given PV system, positive TSO regulation means a negative deviation from the forecast. Based on this, the totals of the negative and positive divergences compared to the actual power generation were calculated from the 60-min or 15-min data for one year for each country. The need for positive regulation by the TSO was marked with a negative sign, since it means a power deficit in the given system, while negative regulation by the TSO was marked with a positive sign, because it refers to an energy surplus in the system. The method of calculation as well as the deviation between the forecast and the actual power generation were illustrated with a Hungarian example (Figure 2). Here the actual production and intraday forecast of a 1196 MWp PV system can be seen.

In the course of the research, the particular 15-min and 60-min data series (intraday forecast, actual production, positive regulation requirement and negative regulation requirement) were analyzed, and within each forecast interval (for Hungary and Poland 15-min, for Slovakia and the Czech Republic 60-min) the difference between the intraday forecast and the actual power generation determined the regulation requirement and the direction of the deviation signaled the direction of regulation. If the difference of the actual electricity production and forecast was a positive one, a negative regulatory requirement emerged, while in the case of a negative difference, a need for positive regulation occurred. Positive and negative regulatory requirements of equaling quantities within a given interval could be balanced by the use of battery-based energy storage systems [87]. Thus, the investigation was searching for solutions for meeting the remaining required power regulation. Its value was obtained after comparing the positive and negative regulatory needs, and summarizing these produced the aggregate regulation requirement.

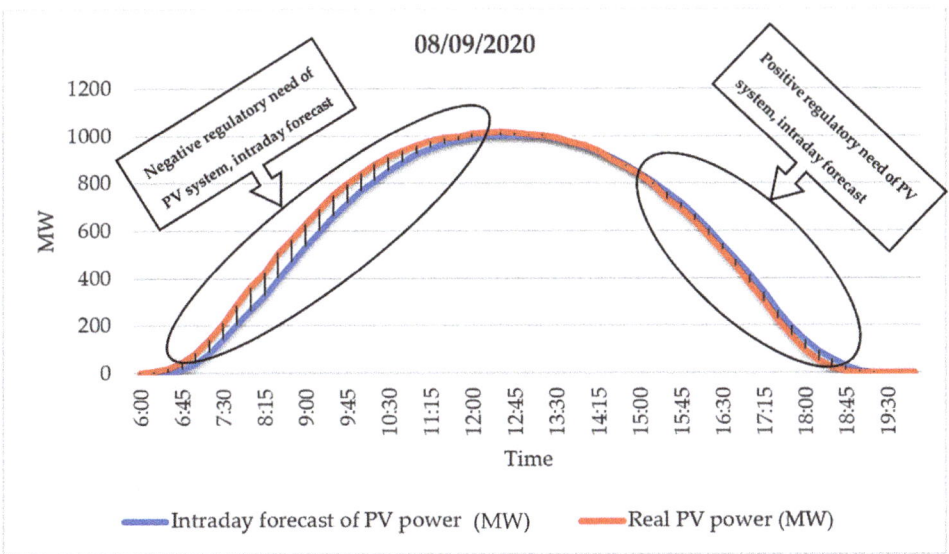

Figure 2. A Hungarian example to illustrate the deviation of real PV power generation and intraday forecast data in the case of a monitored 1196 MWp PV system, 8 September 2020 [81].

The summarized remaining amounts of the needs for negative and positive regulation by the TSO relative to the intraday forecasts were established for all four examined countries. The results show whether the monitored PV systems cause negative or positive required power regulation for the given country. As this paper investigated the application potentials of the power-to-gas technology, the focus was on the countries where the examined PV systems caused negative required power regulation for the TSO.

Since in the case of Poland no data were available in the database of the ENTSO-E prior to May 2020, daily averages were first calculated from the actual production and forecast data of the four available months; then, the annual values were obtained by proportioning these.

3. Results

3.1. Necessary Regulation in the Monitored PV Capacities in the Visegrad Countries

First, those countries were examined where only 60-min data were available, i.e., the Czech Republic and Slovakia. Figure 3 clearly shows that Slovakia has a considerably larger need for regulation per PV unit than the Czech Republic. In the case of the Slovak Republic, this means +4.9 GWh and −6.6 GWh for a PV capacity of 100 MWp, while these figures for the Czech Republic are +0.9 GWh and −1.7 GWh (Figure 3). It needs to be mentioned here, based on Table 2 above, that the PV capacity monitored by the TSO is significantly smaller in Slovakia than in the Czech Republic.

After the countries with the 60-min forecasts, those with the 15-min ones, i.e., Hungary and Poland, were examined. In the case of Hungary, the TSO requires +11 GWh and −7.4 GWh for the regulation of a PV capacity of 100 MWp, while in Poland the amounts needed are only +1.8 GWh and −4.4 GWh (Figure 3). It can be concluded that Hungary needs a lot more regulation for a PV capacity of 100 MWp. It is also to be noted that, according to the figures in Table 2, Poland has a much greater PV capacity monitored by the TSO.

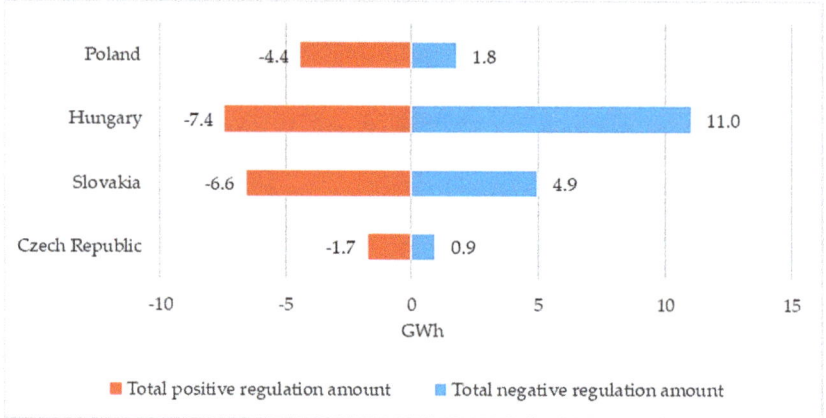

Figure 3. The total regulation amounts in the Visegrad countries necessary for 100 MWp PV systems in the period of 1 September 2019 to 31 August 2020.

Having considered the regulation balances of the countries with the 60-min forecasts, on the basis of the capacity monitored in the examined period, it was concluded that, concerning PV power, both the Czech Republic and Slovakia require positive regulation, i.e., their TSOs need extra energy to keep the forecasts. The necessary extra power for such positive regulation is typically provided by natural gas power plants. It is assumed that the rest of the positive and negative regulation needs that are equal in their absolute values is satisfied by using electrochemical storage in the two countries (disregarding any storage losses). The positive required energy regulation for a PV capacity of 100 MWp in the Czech Republic is altogether 0.8 MWh, while in the case of the Slovak Republic it is 1.6 MWh (Figure 4).

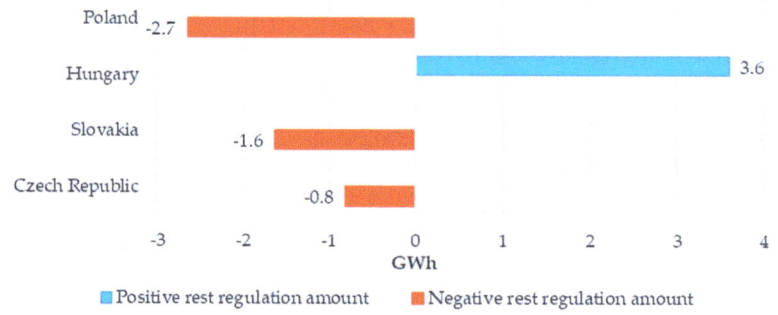

Figure 4. The balance of the negative and positive required energy regulation in the Visegrad countries in the period of 1 September 2019 to 31 August 2020.

While examining the countries that prepare 15-min forecasts, it was found that, concerning their required energy regulation, Hungary requires negative and Poland positive regulation. For a PV capacity of 100 MWp Hungary had 3.6 GWh negative required energy regulation, while Poland had 2.7 GWh positive required energy regulation (Figure 4). It follows from all this that, in the case of the monitored Polish PV capacities, keeping the forecasts requires, apart from electrochemical storage (which is capable of feeding the stored energy into the grid with minor losses, if needed), also natural gas power plants. Conversely, what is needed in Hungary is temporary consumers capable of negative regulation, such as power-to-gas plants.

In the case when a country's positive regulation need equals its negative regulation need, it could manage its regulation by using electrochemical storage. If it wishes to use the stored energy within six hours, the application of Li-ion batteries is recommended, but if the time frame is 6–12 h, NaS or flow batteries (e.g., vanadium redox flow battery) can be the right solution.

As a conclusion, it was found that among the four researched countries, it was only Hungary where negative required energy regulation occurred, i.e., regarding the monitored PV systems only this country had surplus power compared to the forecasts. It is, of course, reasonable to utilize this negative required energy regulation, and power-to-gas technology offers a great solution for this, as it converts surplus electricity into gas. Hungary possesses a well-developed natural gas network, so the transportation of the gas thusly produced could be easily done too.

3.2. A Power-to-Gas Case Study of Hungary

According to the findings above, in Hungary approximately 3.6 GWh negative required energy regulation remains as a result of the regulation of a PV capacity of 100 MWp, if the negative and positive regulation is done by electrochemical technologies where possible. The quantity of the biomethane that could be produced with the help of 3.6 GWh negative required energy regulation in Hungary is shown in Table 3, in the case of PEME and biological methanation, assuming the use of the technology of the Power to Gas Hungary Ltd (Budapest, Hungary).

Table 3. Characteristics of the gas produced by the power-to-gas procedure.

Characteristics	Dimension	Value
Monitored PV in Hungary	MWp	1129
Surplus energy (rest from the regulation)	MWh/100 MWp	3600
Total surplus energy in Hungary for 1129 MWp monitored PV system	MWh	40,644
Produced biomethane	MJ	79,256
Produced biomethane	Nm^3	2,085,679

As it is seen in Table 3, 2,085,679 Nm^3 of biomethane could be produced with the help of 40.6 GWh negative required energy regulation in Hungary. (The calculations were based on the performance data of Hungary's only existing power-to-gas plant, which can produce 97.5 GJ biomethane using 0.1 MW of electrical power by biological methanation and PEME [88]). Based on the Hungarian PV capacities monitored in the ENTSO-E, the above amount of electric energy would have been necessary for the monitored PV systems to follow the forecasts efficiently already in 2020.

Under the Hungarian regulations, there is no such thing as biomethane that is transported by pipeline. Thus, in the calculations the net 0.34 EUR/Nm^3 retail price of natural gas was used, which is paid by a metered customer with consumption below 20 m^3/h at an exchange rate of 350 HUF/EUR. As a result, it was found that natural gas used for keeping the forecasts in Hungary in a value of EUR 709,131 could be replaced regarding the monitored PV capacities.

However, Hungarian agriculture also produces a considerable amount of byproduct, which is not utilized entirely. Some of this is used for making biogas but not with the necessary efficiency, so the methane content of the gas is low, typically only around 40%. For combustion in a gas engine, a methane content of approximately 60% is normally needed. Because of its high methane concentration (approx. 98%), the biomethane produced using the power-to-gas technology is suitable for improving (and thusly preparing for use) the biogas from agriculture, which is usually of poor quality, i.e., with a low methane content.

The approximately 2.1 million Nm^3 biomethane (Table 3) that could be produced potentially in Hungary could be used to enrich about 4 million Nm^3 biomethane that only has a 40% methane content to a 60% methane content. Thus, the 4 million Nm^3 agricultural or landfill gas, so far unused or commonly flared, could be converted from waste into a byproduct, since the utilization of this amount of biogas would be also possible in the future. This means the enrichment and improvement of the end product of approximately 4–6 average-size biogas plants producing low-quality gas to a useful or marketable quality.

4. Conclusions

This paper examined the actual PV power generation and the corresponding intraday forecasts in the four Visegrad countries recorded in the ENTSO-E database and monitored by the regionally responsible TSOs. In the case of the Czech Republic and Slovakia, 60-min, while in the case of Hungary and Poland, 15-min forecast data were available. For better comparability, the need for regulation was calculated for a PV system of a capacity of 100 MWp. After examining the PV capacities monitored by the individual countries' TSOs, it was found that besides the positive and negative regulations of equal absolute values (which are best balanced by the use of electrochemical energy storage), there is negative required energy regulation for the PV systems only in the case of Hungary with an energy amount of 3.6 GWh. In the case of the other three nations (Poland, Slovakia and the Czech Republic) keeping the forecasts of the monitored PV capacities required positive power regulation, which can be done by using natural gas power plants primarily. All this means that, in Hungary, the inclusion of a flexible consumer in the required power regulation process is necessary for keeping the forecasts. In the authors' opinion, this consumer could be a power-to-gas plant in an optimal case, and it could even produce up to 2.1 million Nm^3 biomethane by using the negative required energy regulation.

Considering that Hungary is rich in low-efficiency (typically around 40%) biogas plants, this 2.1 million Nm^3 biomethane could improve and make fit for utilization the end product of 4–6 average-size biogas plants by enriching it to nearly 60%.

The investigation clearly pointed out that, among the countries of the Visegrad Four, it is Hungary where it is worthwhile and necessary to link the existing PV capacities to power-to-gas plants. The results, of course, do not mean that in the other three Visegrad countries (Slovakia, Poland and the Czech Republic) it is impossible to establish power-to-gas plants; they simply indicate that for the Hungarian PV capacities recorded in the ENTSO-E system and monitored by the local TSOs there is more need for power-to-gas plants than in the other countries.

Author Contributions: G.P. conceived, designed and performed the main experiments. The author has read and agreed to the published version of the manuscript.

Funding: This research was funded by Szechenyi 2020 under the EFOP-3.6.1-16-2016-00015.

Conflicts of Interest: The author declares no conflict of interest.

Abbreviations

The following abbreviations are used in this manuscript:

a-Si	Amorphous silicon
EU	European Union
ENTSO-E	European Network of Transmission System Operators for Electricity
EPS	Electric Power System
HREM	Hybrid Renewable Energy Microgrid
m-Si	Monocrystalline silicon
NaS	Natrium-Sulfur
PV	Photovoltaic
PEME	Polymer Electrolyte Membrane Electrolysis
p-Si	Polycrystalline silicon
PVGIS	JRC Photovoltaic Geographical Information System
TSO	Transmission System Operator
VRE	Variable renewable energy

References

1. SolarPower Europe. *Global Market Outlook for Solar Power*; SolarPower Europe: Brussels, Belgium, 2019.
2. REN21. Renewable Energy Policy Network for the 21st Century. In *Renewables 2019 Global Status Report—REN21*; REN21: Paris, France, 2019.
3. Rodríguez, R.A.; Becker, S.; Andresen, G.B.; Heide, D.; Greiner, M. Transmission needs across a fully renewable European power system. *Renew. Energy* **2014**, *63*, 467–476. [CrossRef]
4. Bertsch, J.; Growitsch, C.; Lorenczik, S.; Nagl, S. Flexibility in Europe's power sector—An additional requirement or an automatic complement? *Energy Econ.* **2016**, *53*, 118–131. [CrossRef]
5. Cho, A. Energy's tricky tradeoffs. *Science* **2010**, *329*, 786–787. [CrossRef] [PubMed]
6. Jacobson, M.Z.; Delucchi, M.A. Providing all global energy with wind, water, and solar power, Part I: Technologies, energy resources, quantities and areas of infrastructure, and materials. *Energy Policy* **2011**, *39*, 1154–1169. [CrossRef]
7. Szabó, D. Solar Technologies: Energy Storage Is Installed by the Utility (Napelemek: Energiatárolókat Telepít a Közműszolgáltató). Available online: https://www.napi.hu/magyar_vallalatok/napelemek_energiatarolokat_telepit_a_kozmuszolgaltato.674558.html (accessed on 28 October 2020).
8. Fülöp, M. The First Domestic Public Energy Storage Unit Is Operating (Működik az első hazai közcélú energiatároló egység). Available online: https://www.villanylap.hu/hirek/4904-mukodik-az-elso-hazai-kozcelu-energiatarolo-egyseg (accessed on 28 October 2020).
9. Cochran, J.; Bird, L.; Heeter, J.; Arent, D.J. *Integrating Variable Renewable Energy in Electric Power Markets: Best Practices from International Experience*; National Renewable Energy Lab.(NREL): Golden, CO, USA, 2012.
10. ENTSO-E. TYNDP 2018—Scenario Report. Available online: https://tyndp.entsoe.eu/tyndp2018/scenario-report (accessed on 28 October 2020).
11. Blanco, H.; Faaij, A. A review at the role of storage in energy systems with a focus on Power to Gas and long-term storage. *Renew. Sustain. Energy Rev.* **2018**, *81*, 1049–1086. [CrossRef]
12. European Commission. Clean Energy for All Europeans; European Commission. Available online: https://ec.europa.eu/energy/topics/energy-strategy/clean-energy-all-europeans_en (accessed on 28 October 2020).
13. National Renewable Energy Laboratory (NREL). *Exploration of High-Penetration Renewable Electricity Futures*; National Renewable Energy Laboratory (NREL): Golden, CO, USA, 2012; Volume 1.
14. National Renewable Energy Laboratory (NREL). *Renewable Electricity Generation and Storage Technologies*; National Renewable Energy Laboratory (NREL): Golden, CO, USA, 2012.
15. Hesse, H.; Schimpe, M.; Kucevic, D.; Jossen, A. Lithium-Ion Battery Storage for the Grid—A Review of Stationary Battery Storage System Design Tailored for Applications in Modern Power Grids. *Energies* **2017**, *10*, 2107. [CrossRef]
16. Schimpe, M.; Piesch, C.; Hesse, H.; Paß, J.; Ritter, S.; Jossen, A. Power Flow Distribution Strategy for Improved Power Electronics Energy Efficiency in Battery Storage Systems: Development and Implementation in a Utility-Scale System. *Energies* **2018**, *11*, 533. [CrossRef]
17. Aneke, M.; Wang, M. Energy storage technologies and real life applications—A state of the art review. *Appl. Energy* **2016**, *179*, 350–377. [CrossRef]
18. Han, X.; Liao, S.; Ai, X.; Yao, W.; Wen, J. Determining the Minimal Power Capacity of Energy Storage to Accommodate Renewable Generation. *Energies* **2017**, *10*, 468. [CrossRef]
19. Zsiborács, H.; Hegedűsné Baranyai, N.; Vincze, A.; Háber, I.; Pintér, G. Economic and Technical Aspects of Flexible Storage Photovoltaic Systems in Europe. *Energies* **2018**, *11*, 1445. [CrossRef]
20. Strbac, G.; Aunedi, M.; Pudjianto, D.; Djapic, P.; Teng, F.; Sturt, A.; Jackravut, D.; Sansom, R.; Yufit, V.; Brandon, N. *Strategic Assessment of the Role and Value of Energy Storage Systems in the UK Low Carbon Energy Future*; Imperial College London: London, UK, 2012.
21. European Commission. *EU Reference Scenario 2016*; European Commission: Brussels, Belgium, 2016.
22. International Hydropower Association. *2017 Key Trends in Hydropower*; International Hydropower Association: London, UK, 2017.
23. International Renewable Energy Agency IRENA. *Renewable Energy Capacity Statistics 2017*; International Renewable Energy Agency IRENA: Abu Dhabi, UAE, 2017.
24. Sandia National Laboratories. *DOE Global Energy Storage Database*. Office of Electricity Delivery & Energy Reliability. Available online: https://www.energystorageexchange.org/ (accessed on 28 October 2019).

25. Union of the Electricity Industry–EURELECTRIC. *Hydro in Europe: Powering Renewables Full Report*; Union of the Electricity Industry: Brussels, Belgium, 2011.
26. Gimeno-Gutiérrez, M.; Lacal-Arántegui, R. Assessment of the European potential for pumped hydropower energy storage based on two existing reservoirs. *Renew. Energy* **2015**, *75*, 856–868. [CrossRef]
27. Mantzos, L.; Matei, N.A.; Mulholland, E.; Rózsai, M.; Tamba, M.; Wiesenthal, T. Joint Research Centre Data Catalogue. JRC-IDEES 2015. Available online: https://data.jrc.ec.europa.eu/dataset/jrc-10110-10001 (accessed on 28 October 2020).
28. PANNON Pro Innovations Ltd. Practical Experiences of PV and Storage Systems n.d. Available online: https://ppis.hu/hu (accessed on 28 October 2019).
29. International Hydropower Association. Pumped Storage Tracking Tool n.d. Available online: https://www.hydropower.org/hydropower-pumped-storage-tool (accessed on 28 October 2019).
30. International Hydropower Association. *Hydropower Status Report 2016*; International Hydropower Association: London, UK, 2016.
31. Gyalai-Korpos, M.; Zentkó, L.; Hegyfalvi, C.; Detzky, G.; Tildy, P.; Hegedűsné Baranyai, N.; Pintér, G.; Zsiborács, H. The Role of Electricity Balancing and Storage: Developing Input Parameters for the European Calculator for Concept Modeling. *Sustainability* **2020**, *12*, 811. [CrossRef]
32. International Renewable Energy Agency. *Electricity Storage and Renewables: Costs and Markets to 2030*; International Renewable Energy Agency: Abu Dhabi, UAE, 2017.
33. Zsiborács, H.; Baranyai, N.H.; Vincze, A.; Zentkó, L.; Birkner, Z.; Máté, K.; Pintér, G. Intermittent Renewable Energy Sources: The Role of Energy Storage in the European Power System of 2040. *Electronics* **2019**, *8*, 729. [CrossRef]
34. Igazságügyi Minisztérium. *Magyar Közlöny, 2019*; évi 222. Szám; Igazságügyi Minisztérium: Budapest, Hungary, 2019.
35. Kordmahaleh, A.A.; Naghashzadegan, M.; Javaherdeh, K.; Khoshgoftar, M. Design of a 25 MWe Solar Thermal Power Plant in Iran with Using Parabolic Trough Collectors and a Two-Tank Molten Salt Storage System. *Int. J. Photoenergy* **2017**, *2017*, 1–11. [CrossRef]
36. Noman, A.M.; Addoweesh, K.E.; Alolah, A.I. Simulation and Practical Implementation of ANFIS-Based MPPT Method for PV Applications Using Isolated Ćuk Converter. *Int. J. Photoenergy* **2017**, *2017*, 3106734. [CrossRef]
37. Daliento, S.; Chouder, A.; Guerriero, P.; Pavan, A.M.; Mellit, A.; Moeini, R.; Tricoli, P. Monitoring, Diagnosis, and Power Forecasting for Photovoltaic Fields: A Review. *Int. J. Photoenergy* **2017**, *2017*, 1356851. [CrossRef]
38. Sefa, İ.; Demirtas, M.; Çolak, İ. Application of one-axis sun tracking system. *Energy Convers. Manag.* **2009**, *50*, 2709–2718. [CrossRef]
39. REN21. *Renewables 2018 Global Status Report—REN21*; REN21: Paris, France, 2018.
40. Nengroo, S.; Kamran, M.; Ali, M.; Kim, D.-H.; Kim, M.-S.; Hussain, A.; Kim, H.; Nengroo, S.H.; Kamran, M.A.; Ali, M.U.; et al. Dual Battery Storage System: An Optimized Strategy for the Utilization of Renewable Photovoltaic Energy in the United Kingdom. *Electronics* **2018**, *7*, 177. [CrossRef]
41. Turner, J.A. A realizable renewable energy future. *Science* **1999**, *285*, 687–689. [CrossRef] [PubMed]
42. Lin, A.; Lu, M.; Sun, P.; Lin, A.; Lu, M.; Sun, P. The Influence of Local Environmental, Economic and Social Variables on the Spatial Distribution of Photovoltaic Applications across China's Urban Areas. *Energies* **2018**, *11*, 1986. [CrossRef]
43. Liu, Z.; Wu, D.; Yu, H.; Ma, W.; Jin, G. Field measurement and numerical simulation of combined solar heating operation modes for domestic buildings based on the Qinghai–Tibetan plateau case. *Energy Build.* **2018**, *167*, 312–321. [CrossRef]
44. Alsafasfeh, M.; Abdel-Qader, I.; Bazuin, B.; Alsafasfeh, Q.; Su, W.; Alsafasfeh, M.; Abdel-Qader, I.; Bazuin, B.; Alsafasfeh, Q.; Su, W. Unsupervised Fault Detection and Analysis for Large Photovoltaic Systems Using Drones and Machine Vision. *Energies* **2018**, *11*, 2252. [CrossRef]
45. Hosenuzzaman, M.; Rahim, N.A.; Selvaraj, J.; Hasanuzzaman, M.; Malek, A.B.M.A.; Nahar, A. Global prospects, progress, policies, and environmental impact of solar photovoltaic power generation. *Renew. Sustain. Energy Rev.* **2015**, *41*, 284–297. [CrossRef]
46. Roth, W. *General Concepts of Photovoltaic Power Supply Systems*; Fraunhofer Institute for Solar Energy Systems ISE: Freiburg, Germany, 2005; pp. 1–23.

47. Kumar Sahu, B. A study on global solar PV energy developments and policies with special focus on the top ten solar PV power producing countries. *Renew. Sustain. Energy Rev.* **2015**, *43*, 621–634. [CrossRef]
48. Elavarasan, R.M. The Motivation for Renewable Energy and its Comparison with Other Energy Sources: A Review. *Eur. J. Sustain. Dev. Res.* **2018**, *3*, em0076. [CrossRef]
49. Zsiborács, H.; Pályi, B.; Pintér, G.; Popp, J.; Balogh, P.; Gabnai, Z.; Pető, K.; Farkas, I.; Baranyai, N.H.; Bai, A. Technical-economic study of cooled crystalline solar modules. *Sol. Energy* **2016**, *140*. [CrossRef]
50. Benick, J.; Richter, A.; Muller, R.; Hauser, H.; Feldmann, F.; Krenckel, P.; Riepe, S.; Schindler, F.; Schubert, M.C.; Hermle, M.; et al. High-Efficiency n-Type HP mc Silicon Solar Cells. *IEEE J. Photovolt.* **2017**, *7*, 1171–1175. [CrossRef]
51. Cosme, I.; Cariou, R.; Chen, W.; Foldyna, M.; Boukhicha, R.; i Cabarrocas, P.R.; Lee, K.D.; Trompoukis, C.; Depauw, V. Lifetime assessment in crystalline silicon: From nanopatterned wafer to ultra-thin crystalline films for solar cells. *Sol. Energy Mater. Sol. Cells* **2015**, *135*, 93–98. [CrossRef]
52. Green, M.A.; Emery, K.; Hishikawa, Y.; Warta, W.; Dunlop, E.D. Solar cell efficiency tables (version 48). *Prog. Photovolt. Res. Appl.* **2016**, *24*, 905–913. [CrossRef]
53. *International Energy Agency Technology Roadmap Solar Photovoltaic Energy*, 2014th ed.; International Energy Agency: Paris, France, 2014; pp. 1–60.
54. Krempasky, J. *Semiconductors, Questions & Answers*; Technical Publishing House: Budapest, Romania, 1977.
55. Panasonic Corporation. *Solar Cell Achieves World's Highest Energy Conversion Efficiency of 25.6% at Research Level*; Panasonic Corporation: Kadoma City, Japan, 2014.
56. Verlinden, P.; Deng, W.; Zhang, X.; Yang, Y.; Xu, J.; Shu, Y.; Quan, P.; Sheng, J.; Zhang, S.; Bao, J. Strategy, development and mass production of high-efficiency crystalline Si PV modules. In Proceedings of the 6th World Conference on Photovoltaic Energy Conversion, Kyoto, Japan, 23–27 November 2014.
57. Zsiborács, H.; Pályi, B.; Baranyai, H.N.; Veszelka, M.; Farkas, I.; Pintér, G. Energy performance of the cooled amorphous silicon photovoltaic (PV) technology. *Idojaras* **2016**, *120*, 415–430.
58. Green, M.A.; Dunlop, E.D.; Levi, D.H.; Hohl-Ebinger, J.; Yoshita, M.; Ho-Baillie, A.W.Y. Solar cell efficiency tables (version 54). *Prog. Photovolt.* **2019**, 1–11. [CrossRef]
59. SecondSol Inc. New and Used PV Module Prices. Available online: https://www.secondsol.com/de/index.htm (accessed on 28 October 2020).
60. Kondo, M.; Yoshida, I.; Saito, K.; Matsumoto, M.; Suezaki, T.; Sai, H.; Matsui, T. Development of Highly Stable and Efficient Amorphous Silicon Based Solar Cells. In Proceedings of the 28th European Photovoltaic Solar Energy Conference and Exhibition, Paris, France, 30 September–4 October 2013; WIP: Villepinte, France, 2013; pp. 2213–2217.
61. ÖKO-HAUS GmbH. Information on the Prices of a-Si Solar Modules, Price Quotation. Available online: https://www.oeko-haus.com/ (accessed on 28 October 2020).
62. Enjavi-Arsanjani, M.; Hirbodi, K.; Yaghoubi, M. Solar Energy Potential and Performance Assessment of CSP Plants in Different Areas of Iran. *Energy Procedia* **2015**, *69*, 2039–2048. [CrossRef]
63. Manoj Kumar, N.; Chopra, S.S.; Chand, A.A.; Elavarasan, R.M.; Shafiullah, G.M. Hybrid renewable energy microgrid for a residential community: A techno-economic and environmental perspective in the context of the SDG7. *Sustainability* **2020**, *12*, 3944. [CrossRef]
64. Statista, I. Cumulative Solar Photovoltaic Capacity Globally as of 2017, by Select Country (in Gigawatts). 2018. Available online: https://www.statista.com/statistics/264629/existing-solar-pv-capacity-worldwide/ (accessed on 28 October 2019).
65. Vokas, G.A.; Zoridis, G.C.; Lagogiannis, K.V. Single and Dual Axis PV Energy Production over Greece: Comparison Between Measured and Predicted Data. *Energy Procedia* **2015**, *74*, 1490–1498. [CrossRef]
66. Eke, R.; Senturk, A. Performance comparison of a double-axis sun tracking versus fixed PV system. *Sol. Energy* **2012**, *86*, 2665–2672. [CrossRef]
67. Solargis.com. Solar Resource Maps and GIS Data for 200+ Countries. 2017. Available online: https://solargis.com/maps-and-gis-data/overview (accessed on 28 October 2020).
68. Bai, A.; Popp, J.; Balogh, P.; Gabnai, Z.; Pályi, B.; Farkas, I.; Pintér, G.; Zsiborács, H. Technical and economic effects of cooling of monocrystalline photovoltaic modules under Hungarian conditions. *Renew. Sustain. Energy Rev.* **2016**, *60*, 1086–1099. [CrossRef]
69. Scarlat, N.; Dallemand, J.F.; Fahl, F. Biogas: Developments and perspectives in Europe. *Renew. Energy* **2018**, *129*, 457–472. [CrossRef]

70. Kampman, B.; Leguijt, C.; Scholten, T.; Tallat-Kelpsaite, J. *Optimal Use of Biogas from Waste Streams: An Assessment of the Potential of Biogas from Digestion in the EU Beyond 2020*; European Commission: Brussels, Belgium, 2017.
71. Siddiqui, S.; Zerhusen, B.; Zehetmeier, M.; Effenberger, M. Distribution of specific greenhouse gas emissions from combined heat-and-power production in agricultural biogas plants. *Biomass Bioenergy* **2020**, *133*, 105443. [CrossRef]
72. Garcia, N.H.; Mattioli, A.; Gil, A.; Frison, N.; Battista, F.; Bolzonella, D. Evaluation of the methane potential of different agricultural and food processing substrates for improved biogas production in rural areas. *Renew. Sustain. Energy Rev.* **2019**, *112*, 1–10. [CrossRef]
73. Götz, M.; Lefebvre, J.; Mörs, F.; McDaniel Koch, A.; Graf, F.; Bajohr, S.; Reimert, R.; Kolb, T. Renewable Power-to-Gas: A technological and economic review. *Renew. Energy* **2016**, *85*, 1371–1390. [CrossRef]
74. Bailera, M.; Lisbona, P.; Romeo, L.M.; Espatolero, S. Power to Gas projects review: Lab, pilot and demo plants for storing renewable energy and CO_2. *Renew. Sustain. Energy Rev.* **2017**, *69*, 292–312. [CrossRef]
75. Martin, M.R.; Fornero, J.J.; Stark, R.; Mets, L.; Angenent, L.T. A single-culture bioprocess of methanothermobacter thermautotrophicus to upgrade digester biogas by CO_2-to-CH_4 conversion with H_2. *Archaea* **2013**, *2013*. [CrossRef] [PubMed]
76. Simonis, B.; Newborough, M. Sizing and operating power-to-gas systems to absorb excess renewable electricity. *Int. J. Hydrogen Energy* **2017**, *42*, 21635–21647. [CrossRef]
77. Ghaib, K.; Ben-Fares, F.Z. Power-to-Methane: A state-of-the-art review. *Renew. Sustain. Energy Rev.* **2018**, *81*, 433–446. [CrossRef]
78. Csedő, Z.; Zavarkó, M. The role of inter-organizational innovation networks as change drivers in commercialization of disruptive technologies: The case of power-to-gas. *Int. J. Sustain. Energy Plan. Manag.* **2020**, *28*, 53–70. [CrossRef]
79. Natural Gas Transmission—Our Businesses—MOLGroup. Available online: https://molgroup.info/en/our-business/natural-gas-transmission/natural-gas-transmission-1 (accessed on 23 October 2020).
80. Hungarian Government. *National Energy Strategy 2030*; Hungarian Government: Budapest, Hungary, 2012.
81. Elia Group Solar-PV Power Generation Data. 2020. Available online: https://www.elia.be/en/grid-data/power-generation/solar-pv-power-generation-data (accessed on 11 September 2020).
82. Hungarian Transmission System Operator—MAVIR ZRt. PV Power Generation, Estimation and Fact Data. 2020. Available online: http://mavir.hu/web/mavir/naptermeles-becsles-es-teny-adatok (accessed on 11 September 2020).
83. European Network of Transmission System Operators for Electricity (ENTSO-E). ENTSO-E Transparency Platform. Available online: https://transparency.entsoe.eu/dashboard/show (accessed on 11 September 2020).
84. Hungarian Transmission System Operator—MAVIR ZRt. Transparency. Available online: https://www.mavir.hu/web/mavir/eromuvi-termeles-forrasok-megoszlasa-szerint-netto-elszamolasi-meresek-alapjan1 (accessed on 11 September 2020).
85. EUR-Lex—32019R0943-EN-EUR-Lex. Available online: https://eur-lex.europa.eu/legal-content/EN/TXT/?uri=CELEX%3A32019R0943 (accessed on 23 October 2020).
86. National Legislation Database, H. 389/2007; (XII. 23.) Government Regulation. Available online: http://njt.hu/cgi_bin/njt_doc.cgi?docid=112846.354226 (accessed on 18 October 2019).
87. Zsiborács, H.; Hegedűsné Baranyai, N.; Zentkó, L.; Mórocz, A.; Pócs, I.; Máté, K. Electricity Market Challenges of Photovoltaic and Energy Storage Technologies in the European Union: Regulatory Challenges and Responses. *Appl. Sci.* **2020**, *10*, 1472. [CrossRef]
88. Csedő, Z.; Sinóros-Szabó, B.; Zavarkó, M. Seasonal Energy Storage Potential Assessment of WWTPs with Power-to-Methane Technology. *Energies* **2020**, *13*, 4973. [CrossRef]

Publisher's Note: MDPI stays neutral with regard to jurisdictional claims in published maps and institutional affiliations.

© 2020 by the author. Licensee MDPI, Basel, Switzerland. This article is an open access article distributed under the terms and conditions of the Creative Commons Attribution (CC BY) license (http://creativecommons.org/licenses/by/4.0/).

Article

Disruption Potential Assessment of the Power-to-Methane Technology

Gábor Pörzse [1], Zoltán Csedő [2,3,*] and Máté Zavarkó [2,3]

[1] Corvinus Innovation Research Center, Corvinus University of Budapest, 1093 Budapest, Hungary; gabor.porzse@uni-corvinus.hu
[2] Department of Management and Organization, Corvinus University of Budapest, 1093 Budapest, Hungary; mate.zavarko@uni-corvinus.hu
[3] Power-to-Gas Hungary Kft, 5000 Szolnok, Hungary
* Correspondence: zoltan.csedo@uni-corvinus.hu

Abstract: Power-to-methane (P2M) technology is expected to have a great impact on the future of the global energy sector. Despite the growing amount of related research, its potential disruptive impact has not been assessed yet. This could significantly influence investment decisions regarding the implementation of the P2M technology. Based on a two-year-long empirical research, the paper focuses on exploring the P2M technology deployment potential in different commercial environments. Results are interpreted within the theoretical framework of disruptiveness. It is concluded that P2M has unique attributes because of renewable gas production, grid balancing, and combined long-term energy storage with decarbonization, which represent substantial innovation. Nevertheless, empirical data suggest that the largest P2M plants can be deployed at industrial facilities where CO_2 can be sourced from flue gas. Therefore, a significant decrease of carbon capture technology related costs could enable the disruption potential of the P2M technology in the future, along with further growth of renewable energy production, decarbonization incentives, and significant support of the regulatory environment.

Keywords: power-to-methane; disruptive technology; seasonal energy storage; decarbonization; innovation

Citation: Pörzse, G.; Csedő, Z.; Zavarkó, M. Disruption Potential Assessment of the Power-to-Methane Technology. *Energies* **2021**, *14*, 2297. https://doi.org/10.3390/en14082297

Academic Editor: Attila R. Imre

Received: 17 March 2021
Accepted: 14 April 2021
Published: 19 April 2021

Publisher's Note: MDPI stays neutral with regard to jurisdictional claims in published maps and institutional affiliations.

Copyright: © 2021 by the authors. Licensee MDPI, Basel, Switzerland. This article is an open access article distributed under the terms and conditions of the Creative Commons Attribution (CC BY) license (https://creativecommons.org/licenses/by/4.0/).

1. Introduction

Novel solutions on renewable energy integration and energy storage challenges [1] are driving the global transformation of the energy sector [1–6], due to a significant increase of solar and wind cumulative capacity [7]. This process will even accelerate within the EU because of the European Green Deal (December 2019) and the European Climate Pact (December 2020 [8]). Power-to-methane (P2M) technologies should be considered among these novel (but already commercially ready) solutions, as P2M is suitable for seasonal energy storage by utilizing capacities of the natural gas grid, as well as grid-balancing and carbon reuse [9–11].

The potential impact of P2M technology on the energy sector has already appeared in the power-to-gas (P2G) literature, which continuously broadens with novel technical and economic studies [12–14]. There is a consensus of the crucial role of the P2G technology for the future energy sector [12,15]. Innovation management aspects of P2G technology have also been already (partly) covered [16,17]. A key term and a key phenomenon, however, in the intersection of these three key topics (1: future impact, 2: techno-economic aspects, 3: innovation), called "disruptive technology" and "disruption" are overlooked in the literature despite its importance. The disruptiveness of a technology is highly important from investment aspects because disruptive technologies usually seem inferior from a certain performance aspect compared to other, better-known solutions, even though, later, they can change the dynamics of a whole sector. In other words, investing in P2M on a company level and/or state level could affect organizational/sectoral competitiveness, as it should enable building new competencies through innovation [18], adaptation to new environmental changes [19], sustained and also sustainable growth [20].

The disruptive technology theory is usually applied in business and innovation management research and is less frequently applied in technical contexts or for examining new energy technologies. In broader context, technology- and disruption-focused scholars currently pay attention to digital solutions (e.g., [21–25]), and there are few similar disruption-focused studies in energy research (e.g., [26–29]). For example, Ullah et al. [30] examined the adoption of blockchain technology for energy management in developing countries discussing the distributed ledger technology as disruptive. Zeng et al. [31] pointed out that conventional energy technologies are dominant in the energy sector, and found that low price, high consistency, and high improvement rate are key for the diffusion of renewable energy technologies. In contrast to this broad view, a narrower approach was followed by Müller and Kunderer [32] when they predicted the potential disruption hazard of redox-flow batteries towards lithium-ion batteries with quantitative methodology. This study focuses on the attributes of P2M and does not determine ex-ante its disruptiveness, nor their competing technologies, but identifies them based on empirical data collection and analysis.

Based on the literature, P2M technology in a macro-economic context means an opportunity not only for seasonal energy storage but for decarbonization, as well, as it converts CO_2 into CH_4 in the presence of H_2 [33]. Moreover, P2M can provide e.g., sector coupling [34], new business opportunities on a company-level [35], or also new challenges for the regulators [16]. Thus, P2M seems to be disruptive at first sight. By definition, however, some other questions arise regarding the disruptiveness of the P2M technology. Even so, according to Christensen et al., who introduced the term disruptive technology in 1995 [36], numerous experts and researchers use wrongly the term "disruptive innovation" because it is not only about shaking up an industry and struggling companies which were formerly market-leading [37]. Consequently, to identify a disruptive technology, it is worth going back to the fundamentals of the theory. The main research question of the paper is whether P2M is a disruptive technology by definition or not. Based on the above, the working hypothesis of the paper is that P2M could become a disruptive technology. Main reasons for the conditional approach are (1) the future time horizon of the examined phenomenon, and (2) the hybrid (quantitative and qualitative) methodology applied.

The main practical contribution of this study is that it provides insights for P2M deployment planning within different technical environments, and it identifies those factors which would incite companies and governments to invest resources into P2M deployment in large-scale. The main theoretical contribution is that, to our best knowledge, this study is the first that uses the fundamentals of the disruptive theory in power-to-gas research. Furthermore, the study is contributing to the P2M research field with its hybrid methodology as well: while solely qualitative studies often cannot utilize findings from the field to make general conclusions, and quantitative studies sometimes overlook underlying aspects and challenges of technology development which could be explored by only qualitative methodology, this study involves both qualitative and quantitative data collection and analyses in order to extend empirical findings to broader conclusions for P2M technology development.

2. Materials and Methods

2.1. Research Framework

The counterpoint of the disruptive technology is the "sustaining technology." The sustaining technologies incorporate incremental developments and fit mainstream customer needs. In contrast, disruptive technologies are wholly new solutions, and they create value with an entirely different attribute package that initially does not meet mainstream needs [37]. Instead, they are viable in a niche or low-end market (which is less profitable), or even on a previously non-existing market which is created by a disruptive technology itself, changing non-consumers to consumers [37]. Thus, an important question is, in case of possible disruptiveness of P2M, (RQ1) what are the key attributes of P2M for potential technology adopters and how can they be evaluated compared to other (maybe sustaining) technologies? Regarding this research question, the focus is on the technical opportunities which outline value creation for technology adopters, as previous research showed that economic conditions of day-to-day P2M operations are highly dependent on state support and regulatory environment [17], which is not easy to predict.

While common examples for disruptive solutions are the Netflix streaming services [37] and the copiers of Canon in the late 1970s [38], one can see that interpreting disruptiveness in case of P2M is more complex than in sectors where there are numerous potential customers like in the entertainment/media or printing industry. Consequently, instead of focusing on possible number of consumers, in case of P2M, the assessment of disruptiveness should focus on possible plant sizes on different sites and their compared cost–benefit ratio. This is in line with recent P2M-specific research of Böhm et al. [39] who found a growing need for multi-MW$_{el}$ plants, as global demand for electrolysis and also for methanation can far exceed 1000 GW$_{el}$. Hence, the second research question is (RQ2) what is the largest P2M plant size possible at different types of sites and what sites are preferred for large-scale P2M deployments as possible low-end and high-end segments? This comparison is relevant also because of the possible public funding deriving from the P2G initiatives of the Hungarian National Energy Strategy 2030 [17,40], which must be distributed to sites with best cost–benefit ratios. This means that capital costs of deployment are also important decision factors besides technical opportunities.

The third research question is based on the theoretical phenomenon of disruptiveness. Over time, changes in the market environment and further developments of a disruptive technology result in higher performance compared to sustaining technologies. Therefore, mainstream customers will choose the disruptive solutions over sustaining technologies [38]. Based on these expected changes, regarding the P2M technology, a relevant question is (RQ3) which environmental factors and technological advancements could lead to superior performance compared to other (maybe sustaining) technologies and accelerate the process of P2M implementation? These change aspects dominantly could cover core technology development, complementary technologies, input or output conditions.

Figure 1 illustrates the research framework which was applied to examine P2M as a potentially disruptive technology and did not fix ex-ante that P2M was a disruptive technology. It means that the research framework has also left space to empirically identify whether the underlying assumption was correct and, if it was, why.

2.2. Research Methodology

In line with the carbon-neutrality strategy of the EU, in Hungary, grid-scale implementation of P2M technology is possible and also promising because of a rapid increase of photovoltaics [41], and a 6,330,000,000 m^3 capacity of the natural gas grid suitable for seasonal energy storage [17]. The authors conducted action research in Hungary, at Power-to-Gas Hungary Kft., from 2018 to 2020. The company operates its P2M prototype since 2018 and plans to implement its innovative biomethanation solution in grid-scale. The action research method, which generally incorporates practical ("social") actions, theoretical examinations, iterations between theory and practice (actions and research) to generate change, and new knowledge [42–44], was in this context dominantly technology- and investment-focused. It means that it contained research for potential P2M sites in Hungary, technical data collection from the potential sites, analysis of the maximum plant size, infrastructural conditions and synergies, on-site consultations, collaborative facility-planning with site operators, and interpreting data according to previous studies and the fundamentals of the disruptive theory presented in the Introduction section.

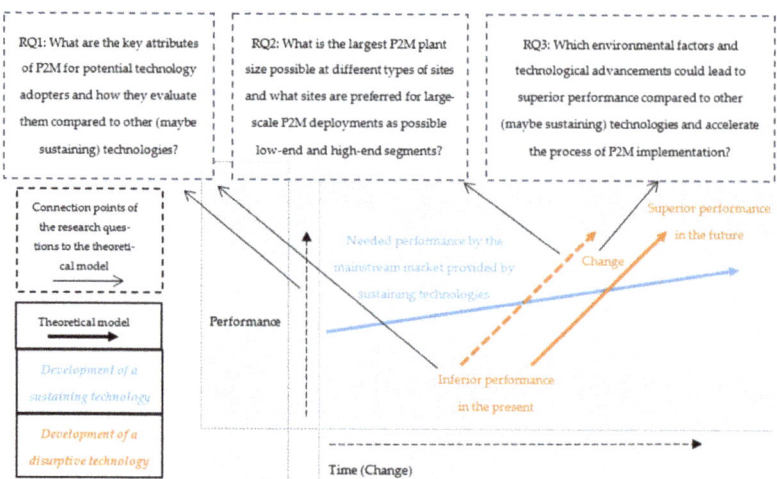

Figure 1. Research framework based on Bower and Christensen's original model [36]. Research question 1 (RQ1) examines whether there are unique attributes of the P2M compared to other technologies from the aspect of potential technology adopters. Research question 2 (RQ2) focuses on the maximum potential of its (maybe unique) attributes and the environments, where these can be realized with the best cost–benefit ratio. Research question 3 (RQ3) deals with the factors which could accelerate the realization of this maximum P2M potential.

2.3. Focal Solution and Its Main Characteristics

The focal solution of this research is the P2M technology of Power-to-Gas Hungary Kft. The prototype of the company includes polymer electrolyte membrane (PEM) electrolysis and biological methanation, the main component of which is a patented, robust, and highly efficient microorganism (Methanothermobacter thermautotrophicus) [45]. This archaea strain is capable of converting over 99% of the CO_2 to methane, resulting in a product gas with methane content above 97% which can be injected into the natural gas grid. The main reactions of the applied P2M process are the following:

- Electrolysis: $4H_2O \rightarrow 4H_2 + 2O_2 + \text{Heat}$;
- Methanation: $CO_2 + 4H_2 \rightarrow CH_4 + 2H_2O$.

Based on the detailed technical description of the three steps of the applied P2M process (electrolysis, methanation, injection) of the Power-to-Gas Hungary Kft. which has been recently published [13], the possible performance aspects of the focal P2M technology and their base data are presented in Table 1.

Table 1. P2M performance aspects and related technical data (based on [13]).

Performance Aspect	Base Data, Description in Case of a 1 MW_{el} Biomethanation Plant
CO_2 input	Ca. 53 CO_2 Nm^3/h.
CH_4 production	Ca. 52 Nm^3/h (ca. 97–98% of the CO_2 input)
Energy storage	No limit, if a connection to the natural gas grid is available
H_2 output (P2H) and input (P2M)	Ca. 212 Nm^3/h (with regard to the ca. 4:1 or 4.1:1 ratio of H_2 and CO_2)
Electricity consumption	Ca. 4.7 kWh / Nm^3 H_2

Starting from the research framework, it is important to analyze what type of electrolyzers could provide hydrogen for the methanation step on a large-scale. While solid-oxide electrolysis (SOEL) is under development for commercialization [46], alkaline electrolysis (AEL) and polymer electrolyte membrane electrolysis (PEMEL) have been implemented in large P2M facilities [14,47,48]. For example, the Audi e-Gas plant has alkaline electrolyzers (6 MW$_{el}$), while the STORE&GO demonstration site at Solothurn has PEM electrolyzers (700 kW$_{el}$) [49]. Regarding the required flexibility because of the volatile renewable electricity production, PEMEL can be started within seconds, while starting AEL may need 1–10 min [50]. Some manufacturers stated, however, that AEL is also capable of rapid reaction and fast warm start [51]. AEL and PEMEL have been implemented in larger sizes until now (e.g., the other two sites of the STORE&GO project in Italy and Germany have 1 and 2 MW AELs [11]).

In the methanation step, two technologies have also been applied on a commercial-scale: the catalytic (chemical) methanation (used e.g., in AUDI e-Gas) and the biological methanation (e.g., Biocat project). While biological methanation applies microorganisms as biocatalysts and needs low pressure and temperature (ca. 60–70 °C), catalytic methanation uses chemical, often nickel- or ruthenium-based catalysts, and sometimes more than 100 °C is needed to reach high CO_2 conversion [52,53]. Based on Frontera et al. [54], the CO_2 conversion rate can be variable (from 50–60% to 80–90% or higher) depending on the type of the chemical catalyst and the temperature. Biological methanation, however, can provide consistency in this sense; at low temperature and low pressure, it could lead to as Table 95% CO_2 conversion rate, with high flexibility for pausing, stopping, and restarting methane production [55,56]. The nutrition of the biocatalysts, however, is key in this case [57], while the efficiency of the whole process could be further improved if waste heat could be utilized, which is generally difficult at this lower temperature [58]. Two other solutions are worth mentioning, which are currently in the development phase. First, the bioelectrochemical system for electromethanogenesis (EMG-BES) may need an even lower temperature for the reaction (ca. 25–35 lied on a commercial-scale: the catalytic (chemical) methanation (used e.g., in AUDI e-Gas) and the biological methanation (e.g., Biocat project). While biological methanation applies microorganisms as biocatalysts and needs low pressure and temperature (ca. 60–70 °C) [59]. Second, in the case of biogas plants, novel in-situ biological biogas upgrading (BGU) with hydrogenotrophic methanogens in a mixed culture can be used for methanation by supplementing H_2 from renewable sources [60].

Disruptiveness of P2M technology would mean large, commercial-scale plants. The focal solution consisting of PEMEL and biological methanation seem applicable for the study. The capital expenditures (CAPEX), however, can be critical, even if state support would be available for the investments. A recent calculation based on the data of the STORE&GO project and additional field research, the CAPEX of a 1 MW$_{el}$ P2M plant using PEMEL and biological methanation would be 4,806,000 EUR in 2025 at a large wastewater treatment plant (WWTP) [13].

2.4. Data Collection and Analysis

The authors collected empirical data from potential sites for P2M implementation to explore:

- sites that may have proper infrastructural, input, and output conditions for a biological methanation plant with a world-leading size (over 1 MW$_{el}$)
- aspects of consumer evaluation about P2M and competing solutions, if there are any
- site-specific factors that would enable the increase of the plant size or the feasibility of a large P2M plant.

From 2018 to 2020, the authors contacted potential sites, among which there were

- agricultural biogas plants (ABPs)
- wastewater treatment plants with biogas plants (WWTPs)
- bioethanol plants (BEPs)
- industrial plants (INPs) with CO2 emission (e.g., power generation, petrochemicals, cement plant).

While in case of ABPs, WWTPs, and BEPs, the CO_2 input for methanation can be provided with an easily and efficiently useable carbon source (the CO_2 content of the biogas and pure CO_2 can be sourced from the exhaust stream of the fermentation during bioethanol production [61]), in the case of INPs CO_2 must be captured from flue gas with Carbon Capture (CC) technologies, for example, at a cement plant [62]. Figure 2 illustrates the input connections of a P2M plant at different sites, showing how CO_2 can be sourced for the methanation phase.

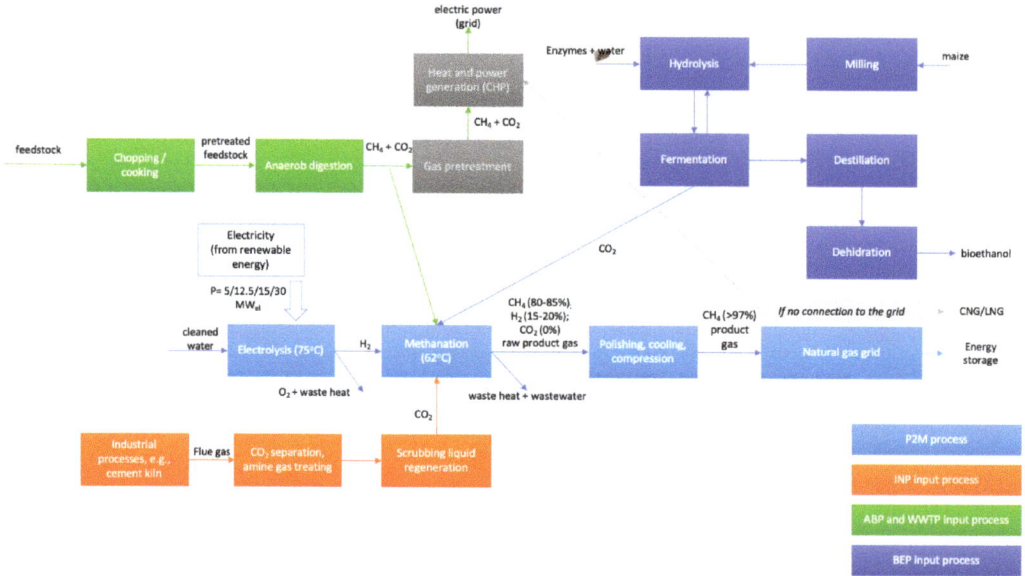

Figure 2. Simplified overview of the P2M process at an INP with carbon capture (orange), at a biogas plant of an agricultural company or a WWTP (green) or at a BEP (navy blue), based on Sinóros-Szabó [63], Laude et al. [61], and Chauvy et al. [62].

The authors collected data from 29 potential sites, which means that at least one personal meeting, teleconference, or online videoconference happened with one or more members of the top management team of the site operator. In these 29 first-round semi-structured interviews at potential sites, the main characteristics of the focal technology were presented to decision-makers, and the infrastructural fit was analyzed on a high-level. After these 29 first-round interviews, second-round semi-structured interviews were focusing on more technical data and included on-site consultations with expert-level operators. This was possible at 14 potential sites based on infrastructural opportunities and availability of the top management.

While the first-round data collection aimed to identify the needs of the potential adopters, the main value-creating attributes of P2M according to these needs and the existence of competing technologies (RQ1), the second-round data collection was aimed to determine the maximum potential of the P2M plant at the focal site and its costs (RQ2), and also to explore the factors of increasing technical and/or financial feasibility of the implementation (RQ3).

Considering these goals, Table 2 summarizes the scope and the methods of data collection and analyzes the efforts undertaken to improve validity, reliability, and generalizability both in qualitative and quantitative sense. The calculations are presented in the Results section.

Table 2. Data collection and data analysis of the research

Data Collection and Analyses	First Round	Second Round
Dominant Methodology	Qualitative	Quantitative
Data source(s)	• Interviews with top management (in person, teleconference, videoconference)	• Interviews with top management (in person or videoconference because of the COVID-19 • Filling technical data form by the site • On-site consultations with expert-level operators
Data collection	• Technology and infrastructure development plans and motivations (needs) behind them • Value creating potential of P2M technology regarding these plans • Other technologies that emerged, as alternatives, and their expected value added	• P2M-specific technical, technological, infrastructural characteristics, such as power supply, water supply, CO_2 input, infrastructural connections, use of byproducts (waste heat, oxygen) • Site-specific technical, technological, infrastructural characteristics, such as fermentation at ABPs and WWTPS, flue gas composition at INPs, CO_2 from fermentation at BEPs
Data analysis	• Grounded theory coding technique (open coding, axial coding, selective coding) [64]	• Technical calculations, scenario-analyses, and cost–benefit assessments
Improving validity	Conducting two-year-long research to explore the research area deeply, reaching theoretical saturation with 44 interviews (first and second round together), similarly to high-quality studies with similar topics and methods (see, e.g., [65]: 17, [66]: 31, [67]: 20 interviews)	Assessing outputs based on technical parameters, but also cost–benefit ratio based on CAPEX
Improving reliability	Always more than one interviewer, involving author as a researcher from outside the power-to-gas area. The second round reinforced and fine-tuned the results of the first round.	Building on calculations of previously published studies and empirical data of the prototype of Power-to-Gas Hungary Kft. including more than 30,000 measurements since 2018 (e.g., CO_2 conversion, the composition of the product gas, the volume of the product gas).
Improving generalizability	Iteration between empirical data and the theory of disruptive technologies	Assessing more types of potential sites, and at least two from each type

3. Results

3.1. Mainstream Needs and Emerged Alternative Solutions at Different Sites

Based on semi-structured interviews with top management teams of different sites, the overall mainstream need is producing and utilizing more renewable energy. While meeting this need, P2M has faced different competing technologies at different sites. Table 3 presents the specific opportunities and competing technologies by the identified valuable attributes for potential adopters in producing and utilizing more renewable energy. The table does not contain every possible technology and every aspect of potential competitive advantages of them because it is built on empirical data from the field, the evaluation aspects of the interviewees, but it was iterated with scientific literature:

(1) In case of biogas plants both in an agricultural environment and at WWTPs, BGU can be considered as a competing technology to produce renewable gas (biomethane). As there were more than 400 facilities with BGU to produce biomethane in 2015 worldwide [68], and even in Hungary there are two [69], one could argue that BGU is a more mature technology than P2M. This higher technology readiness level (TRL) that is associated with lower risks seemed to be an important factor for decision-makers, as prudent risk management appeared as a strategic task, for example in the case of WWTPs [13]. Regarding the other elements of the attribute package, focal

(2) P2M technology with a separate reactor and the patented archaea could have a higher decarbonization effect, as some BGU technologies do not involve CO_2 conversion (only separation) and even if H_2 is injected to be reacted with endogenous CO_2 to produce CH_4 during in situ biological upgrading, the average CO_2 removal rate is varying between 43–100%, depending on reactor type and substrate [68]. Furthermore, a clean archaea culture could provide more flexibility for utilizing H_2 from renewable sources than in situ biological BGU based on the rapid shifts between operation modes of the focal solution based on prototype data of Power-to-Gas Hungary Kft. [13].

(2) In case of industrial companies emitting CO_2 that could be used with P2M to produce renewable or low-carbon gas depending on the source of input factors [70], power-to-liquid technology (P2L) emerged as an alternative technology. P2L has also a high potential in the future energy sector [71], especially for transportation, but the plan for the first commercial-scale P2L plant is only recently published [72].

(3) The first phase of renewable methane production, power-to-hydrogen (P2H) can be a standalone solution as well. As presented before, the fast warm start of PEMEL or AEL can be useful for providing grid-balancing services for network operators [50,51]. Even though it means that producing renewable energy (gas) and grid-balancing can be achieved with decreased CAPEX, adding the methanation step with a biocatalyst could also provide flexibility, not only in terms of methane production (avoiding the need for the challenging high volume hydrogen storage [73]), but also by assuring market-flexibility. Market-flexibility means here the opportunity to switch between end-products (hydrogen and methane) according to their market demand. From an operational point of view, adding the methanation step and assuring this market flexibility would lead not only to higher CAPEX, but lower energy efficiency for the whole process as well. The reason for that is the additional conversion step (and the energy demand which might be required for CO_2 capture). Consequently, the value of this market flexibility is highly dependent on the operational context. For example, if (1) hydrogen injection to the natural gas grid remains still strongly limited and/or its local demand is low, but (2) high feed-in-tariffs incite green methane production and/or high carbon taxes incite avoiding carbon emissions, this market-flexibility could add significant value.

(4) Based on the empirical data, if the sites would plan to deploy a large solar park for renewable electricity production, battery energy storage systems (BESS) emerged as a viable option. (In this research, mostly INPs, ABPs, and BEPs have mentioned this option, while some WWTPs stated that they did not have enough free territory to deploy a large solar park.) The main advantages of BESS related to on-site energy storage are the fast response, geographical independence, other energy management functions [74], and also the grid-balancing services [75]. While BESS efficiency for the short-term can be higher than the focal solution's (55–60%) [13], P2M could provide sector coupling and seasonal energy storage which could be valued or supported by state administration as it appeared as an important goal in the Hungarian National Energy Strategy 2030 [18].

(5) Finally, regarding direct decarbonization, Carbon Capture technologies can be relevant. For example, post-combustion capture using wet scrubbing with aqueous amine solutions is commercially advanced [76], but pre-combustion, oxy-fuel combustion and chemical looping combustion are also promising to capture CO_2 from flue gas [77] that a P2M solution is not capable solely (in contrast to biogas which also contains CO_2 and can be injected to the P2M bioreactor). P2M, however, could utilize CO_2 for renewable energy production.

Table 3. P2M attribute package and alternative technologies based on the evaluation of potential adopters iterated with previous studies.

P2M Attribute Package	Competing Technologies	Relevant Sites Based on Empirical Data	Main Advantage of P2M	Main Advantage of Competing Technology
Producing renewable gas or another energy carrier different from electricity	BGU, CO_2 removal or conversion by mixed culture with hydrogenotrophic methanogens	ABPs WWTPs	Higher CO_2 conversion and technical flexibility	Higher TRL
	Power-to-Liquid (P2L)	INPs	Higher TRL	Applicability for another sector (transportation)
Providing grid balancing services	Solely power-to-hydrogen (P2H)	INPs	Market-flexibility	Smaller CAPEX for producing renewable energy and providing flexibility
Short-term and long-term energy storage	Battery energy storage systems (BESS)	INPs ABPs BEPs	Applicability for sector coupling and long-term energy storage	Higher efficiency for short-term energy storage
Direct decarbonization	Carbon Capture (CC) technologies	INPs	CO_2 reuse	Serving decarbonization efforts in case of flue gas, as well

Based on the presented iteration of empirical data and former studies, four main findings can be outlined:

1. There is no other technology that has the same attribute package as P2M (producing renewable energy, providing grid-balancing services, energy storage, and decarbonization).
2. The most unique attribute in the P2M package is the capability for long-term energy storage with CO_2 reuse. Renewable gas production is possible with BGU, as well, or P2L is suitable for sector coupling (renewable energy production with transportation), it also assures market flexibility (hydrogen or hydrocarbon fuel production) and direct decarbonization effect, but not with long-term (seasonal) energy storage. In contrast of BGU and P2L, the maturity of P2M is also favorable: the technology is newer than BGU, and it has been implemented in grid-scale, unlike P2L.
3. The least unique attribute of P2M is providing grid-balancing services because P2H and battery energy storage systems are also similarly capable to provide this short-term flexibility.
4. The listed alternative technologies may compete with P2M in one dimension of the value creation, but they can be complementary solutions not only at national energy system-level but also in a given case of a potential technology adopter. For example, battery energy storage and P2M can be combined for short-term and long-term energy storage. Carbon Capture could also provide the main input (CO_2) for methanation. Similarly, P2H is inevitable for P2M if seasonal energy storage is considered (because electrolysis is the first step to absorb surplus renewable electricity), even though they may compete in renewable gas production or grid-balancing.

In summary, based on potential adopter evaluation of P2M and its potential competitor technologies, the parallel function of decarbonization and seasonal energy storage is the unique element of the P2M attribute package. It is important to highlight that this uniqueness derives from the absolute capability for decarbonization and energy storage, not from a superior performance, compared to e.g., P2H regarding decarbonization. Moreover, this uniqueness is interpreted as a value-creating attribute on a micro-level for a single technology adopter. This is relevant because different approaches could lead to different results regarding performance evaluations in certain dimensions. For example, Zhang et al. [78] showed that P2M for mobility could result higher GHG emissions than conventional natural gas. This is due to an evaluation that includes a system extension, which also reflects the reduced emissions by CC. Nevertheless, these are meso- or macro-level findings, while this disruption-focused paper is concerned about the aspects of single operators which value that their unwanted CO_2 can be converted into methane. Finally, it should be mentioned that competitor technologies in one value-creating dimension are rather complementary solutions if we take a holistic view on all value-creating dimensions.

3.2. Potential Sites for Large-Scale P2M Deployment

From the 14 potential sites of the second-round data collection and analysis, the authors identified those sites where the largest P2M plant could be deployed with biological methanation. The potential plant size can be determined based on the CO_2 input with regard to the stoichiometric ratio of hydrogen and carbon dioxide (4.1:1). Consequently, the maximum electrolyzer capacity (as the indicator of plant size) of a P2M facility is calculated with the presumption of the 4.7 kWh electrical energy demand (see Table 1) for the yield of 1 Nm^3 of hydrogen is 4.7 kWh/Nm [13]^3:

$$P_{P2Mmax} = \dot{V}_{H_2} \cdot 4.7 \frac{kWh}{Nm^3} = \dot{V}_{CO_{2max}} \cdot 4.1 \cdot 4.7 \frac{kWh}{Nm^3} \quad (1)$$

Equation (1) shows that the maximum size of the P2M plant (P_{P2Mmax}) can be estimated based on the electrical energy demand (4.7 kWh/Nm^3) and hydrogen gas volumetric flow (Nm^3/h) ($\dot{V}_{H_2}$) that is calculated by multiplying the maximum CO_2 input (Nm^3/h) ($\dot{V})_{CO_{2max}}$ with its stoichiometric ratio to hydrogen (4.1).

Table 4 shows the largest possible plants by site type based on empirical data collection and the presented equation based on the characteristics of the focal technology. Because of practical reasons, the calculation considered the autonomous development plans of the sites for the next 2–3 years. For example, a biogas plant planned to expand its biogas producing capacities that would result in higher possible P2M plant size.

Table 4. Largest possible P2M plants by site type based on empirical data collection (with rounding because of confidentiality).

	$(\dot{V})_{CO_{2max}}$ Max. Monthly Average CO_2 Input (ca. Nm^3/h)	P_{P2Mmax} Max. Plant Size (ca. MWel)
ABP	700	12.5
BEP	850	15
WWTP	300	5
INP	1650	30

Based on these empirical data and theoretical calculations, the largest P2M plant could be deployed at INPs. Two additional factors, however, should be considered:

1. First, some seasonality could be seen on yearly data of CO_2 production. At certain sites, CO_2 input can be 30–50% lower in certain months than the maximum monthly average. For example, in case of some WWTPs and ABPs, the beginning and the end of the year has lower volumes of biogas production; consequently, there is less CO_2 available to be converted into methane. This phenomenon may lead to a need for balancing renewable energy gas production (and seasonal energy storage) and decarbonization: while from the decarbonization aspect, it would be important to convert as much CO_2 to methane as possible, seasonality in CO_2 emissions limits the financial attractiveness of scaling the plant size up to the maximum emission level.
2. Second, in case of ABPs, BEPs, and WWTPs, CO_2 is available for efficient use within the P2M plant, but in case of INPs (where the largest P2M plants could be deployed),

there is need for carbon capture (CC) technologies as well, in order to separate CO_2 from the flue gas. CC would increase technical complexity, capital, and operational expenditures as well.

3.3. Performance Potential of Large-Scale P2M Plants at Different Sites

Based on these empirical findings, cost–benefit ratios of P2M plants have been assessed according to decarbonization and renewable gas production (as a prerequisite to long-term, seasonal energy storage) at the largest possible plant size, based on the CO_2 source. Even though the deployment of such large P2M plants may not be financially attractive for a single technology adopter, following the decarbonization efforts, it is worth examining what is the performance potential at different sites regarding not only the methane production but the CO_2 reuse. As mentioned in the Introduction section, these comparisons may orient public funding decisions to facilitate decarbonization and seasonal energy storage [13,41]. As P2M deployment requires significant investments, the socio-economic value creation at these sites may influence the location, the number and the size of P2M plants that will be deployed. From a disruptive point of view, these comparisons can outline low-end and high-end segments of the P2M technology.

Figure 3 shows the unit cost of CO_2 reuse by CAPEX at different sites with the largest possible P2M plant size. In line with the origins of the disruptive theory, the model focuses on the factors that can be affected by technology developers (companies), which means that the important, but hardly predictable regulatory-related interventions (e.g., carbon tax) are out of scope. The unit cost is calculated based on the following factors:

- CAPEX of the P2M plant is based on a recent study by Böhm et al. [40], which focuses on future large-scale P2G technology implementations and takes into account the scaling effects as well. Accordingly, cost reductions due to scaling up differ by site types. Current calculations are predictions for 2025 based on the data of 2020 and estimations for 2030 of Böhm et al. Appendix A presents the basis of CAPEX calculations.
- In case of INP, CC technologies would mean additional costs. It was predicted at ca. 40 EUR/tCO_2 (49 USD/tCO_2) for 2025 by Fan et al. [79].
- CO_2 conversion and CH_4 production has been determined based on the prototype data of the Power-to-Gas Hungary Kft with the focal technology. In line with a former study [13], the 1 MW_{el} base case would mean the conversion of 848 tCO_2 and 4.363 MWh CH_4 yearly.
- The ratio of CAPEX and the converted CO_2 and the produced CH_4 is calculated for 20 year-long operations of the plant, with 8000 h operations per year. Detailed data can be seen in Appendix A.

Figure 3 shows that P2M is capable of the best performance at a BEP with 15 MW_{el} P2M potential regarding decarbonization and renewable gas production, due to the scaling effects and the efficiently useable carbon source (no need for CC), and the worst in case of INP where the cost of CC weakens the cost–benefit ratio more than scaling effects improve it. Nevertheless, as previously mentioned, the uniqueness of P2M derives from the capability to provide seasonal energy storage and sector coupling with parallel decarbonization. Consequently, it is important whether infrastructural connections to the natural gas grid are available at (1) these site types and (2) the sites where the largest P2M plants could be deployed. Results showed that, while WWTPs (where smaller plants could be deployed), mostly have a nearby connection to the natural gas grid, this is less frequent in the case of ABPs and BEPs in Hungary. For example, at a BEP with the largest P2M potential, the nearest connection point is 5 km away, while at a ABP with the largest P2M potential, it is 10 km away, where the produced biomethane could be injected into the natural gas grid. Building these missing infrastructural connections would significantly decrease the financial feasibility of P2M seasonal energy storage. If CC technologies would be available at INPs, connection points to the natural gas grid would be more favorable. Consequently,

CC technology associated costs could be an accelerating factor for seasonal energy storage and decarbonization by large-scale P2M plants.

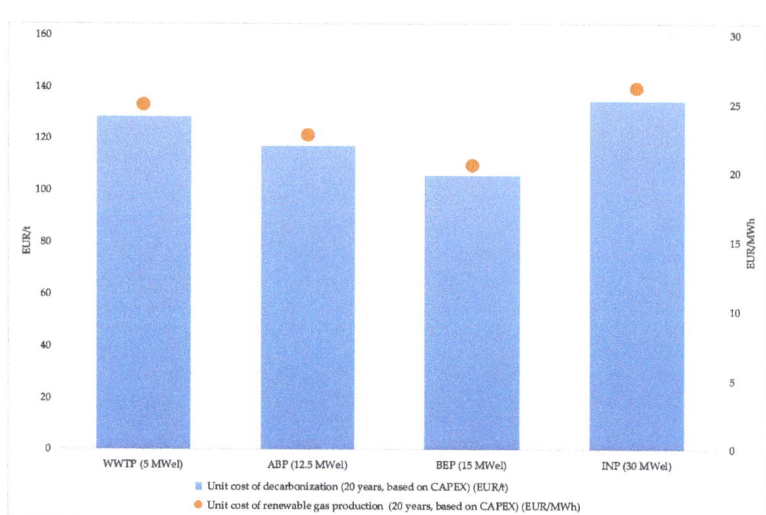

Figure 3. Unit capital cost of decarbonization and renewable gas production (as a prerequisite of seasonal energy storage) of large-scale P2M plants at different sites during their operation (2025–2045).

3.4. Scenarios for 2025 and 2030 Regarding Carbon Capture Cost Reduction

To examine how forecasted CC cost reductions might increase the cost–benefit ratio of large-scale P2M deployments at INPs compared to ABPs or BEPs, 3-3 scenarios have been built for 2025 and 2030. Moreover, further scenarios have been built on the prediction of the P2M CAPEX following Böhm et al. [40] (see Appendix A). Scenarios S1–S3 are differing regarding CC costs as well (Table 5), which were estimated by

- following Fan et al. [79] for the 2025 and 2030 values (S1, ca. 40 EUR/tCO$_2$ in 2025, indicated as 100%; ca. 32 EUR/tCO$_2$ in 2030);
- following Wilberforce et al. [80] showing that CC costs can be around 25 USD/tCO$_2$ mainly at integrated gasification combined cycle (IGCC) and pulverized coal (PC) plants, but also at natural gas combined cycle (NGCC) plants. This is a more optimistic scenario with its 50% reduction (S2), meaning 20 EUR/tCO$_2$ in 2025 and 16 EUR/tCO$_2$ in 2030;
- generating an own scenario to identify the CC cost level which could trigger decision-makers to choose industrial sites with the necessity of CC to deploy a large-scale P2M plant there (S3). For this, another 50% cost reduction is determined.

Table 5. Scenarios based on different carbon capture cost-levels for 2025 and 2030.

CC Cost	Cost Reductions by Scenario	2025	2030	Source
Scenario 1 (S1)	-	40 EUR/tCO$_2$	32 EUR/tCO$_2$	based on Fan et al. [79]
Scenario 2 (S2)	−50%	20 EUR/tCO$_2$	16 EUR/tCO$_2$	based on Wilberforce et al. [80]
Scenario 3 (S3)	−50%	10 EUR/tCO$_2$	8 EUR/tCO$_2$	Own estimation

Figure 4 shows how site preference would change if CC costs would fall by 50% twice. In case of the lines which indicate the cost–benefit ratios of different comparisons, 100% means the performance of the ABP/BEP/WWTP regarding the unit costs presented in the previous chapter and their value in 2030. If the unit cost of decarbonization and seasonal energy storage is lower in the case of 30 MW$_{el}$ P2M+CC configuration at an INP than the value of 5/12.5/5 MW$_{el}$ P2M configurations at a(n) ABP/BEP/WWTP, it means that the cost–benefit ratio of P2M+CC is higher than theirs, so the indicating line goes beyond 100%. Regarding CC costs, 100% means 40 EUR/tCO$_2$ in line with Table 5.

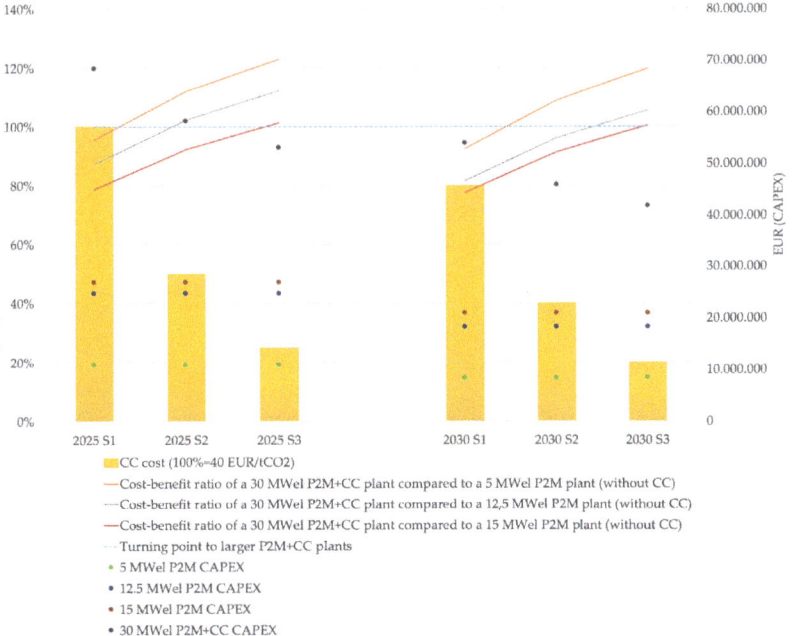

Figure 4. The role of carbon capture costs in choosing certain sites and plant sizes for decarbonization and renewable gas production with P2M.

Findings suggest that the cost–benefit ratio of 30 MW$_{el}$ INPs would be better regarding renewable gas production, seasonal energy storage, and decarbonization at CC costs 20–30 EUR/tCO$_2$ or lower than 5–15 MW$_{el}$ P2M facilities at ABPs or BEPs even if they would have a connection to the natural gas grid in 2025. Nevertheless, due to the estimated cost reductions of P2M CAPEX for 2030, a 12.5 MW$_{el}$ P2M plant would have a better cost–benefit ratio at 16 EUR/tCO$_2$ CC cost than a 30 MW$_{el}$ P2M + CC configuration plant. Assuming that the main goal is seasonal energy storage and the connection to the grid is not an obstacle, 15 MW$_{el}$ or larger P2M facilities at ABPs or BEPs would be competitive with P2M facilities at INPs for decarbonization and seasonal energy storage (renewable gas production and injecting it into the grid) even if costs of CC would radically decrease. In addition, it is important to highlight that, as CC cost would start to decrease from 40 EUR/tCO$_2$, a 30 MW$_{el}$ P2M + CC plant would outperform 5MW$_{el}$ or smaller P2M plants based on the unit capital costs of decarbonization and renewable gas production.

4. Discussion

RQ1 was focusing on key attributes of P2M for potential technology adopters and their evaluation compared to other technologies. According to the literature, disruptive technologies create value with a different attribute package than sustaining technologies, and initially do not meet the mainstream needs. To justify this assumption for P2M, it must be identified whether there are sustaining and disruptive technologies in this market segment at all. As sustaining technologies mean continuous incremental improvements in satisfying mainstream needs, it assumes technologies with widespread utilization and high TRL. Regarding the identified mainstream needs of potential P2M adopters (producing and utilizing more renewable energy) and the recent literature about the identified alternative technologies, mainly BGU and BESS could be considered as sustaining technologies. In the case of BGU and BESS, frequent use and relatively high TRL can be seen [81,82] for renewable energy production and utilization, but there are also novel ways for BGU (TRL3-7) [83] and there are also efforts to optimize and develop the efficiency of batteries [84], which may indicate incremental developments. In contrast, P2H, CC, and P2L are rather in the demonstration phase or less frequently used. In the case of P2H, while low-temperature electrolyzers are at TRL9 (readiness for full-scale implementation), high-temperature electrolysis processes are at TRL6-8 [85]. A recent study, however, pointed out that "the scale of P2H pilots is very small" ([86], p. 1369), and these are demonstration projects, even if one reaches 100 MW (Hybridge). Regarding CC, there are several technologies from TRL2-3, (such as oxygen transport membranes which integrate O_2 separation and combustion) to TRL8-9 (e.g., the commercial CO_2 capture plant in Canada, the Boundary Dam project) [87]. Finally, as there are only plans for P2L facilities on commercial-scale [72] and the P2L technology is rather in demonstration phase with TRL-5-6 [88], P2L cannot be considered as a sustaining technology.

Based on the above, one could argue that P2M can be disruptive against BGU and BESS. This statement can be justified based on the P2M unique attribute package (producing renewable energy, providing grid-balancing services, energy storage, and decarbonization), which is different from BGU and BESS. While BGU is less flexible to provide grid-balancing, BESS does not produce renewable energy. However, it can be also seen that the initial performance of the P2M is inferior compared to them. For example, the capital costs of traditional BGU technologies can be lower, where there is no need for electrolyzers to generate hydrogen [89]. Furthermore, Lithium-Ion Batteries (LIBs) can provide 95–98% efficiency [90]. Assuming that the mainstream market need naturally integrates cost-efficient renewable gas production and high-efficiency energy storage (on the short-term) at ABPs, BEPs, WWTPs, or INPs, P2M has the disruption potential because of this inferiority. Nevertheless, according to the theory, this inferiority of P2M will turn into superior performance later due to the fit of the unique attribute package and environmental changes. Regarding the growing share of renewables in the energy mix, the volatile production may go beyond the capacities of BESSs, and long-term, high volume, seasonal energy storage will be needed. The incitement of this may result in better business opportunities (e.g., high biomethane feed-in-tariff) due to state interventions [18]. This would justify the investments into more CAPEX intensive projects with P2H and P2M (compared to traditional BGU) or expanding the battery-dominated energy storage systems with P2M to realize profits from low priced surplus electricity. As the empirical research pointed out based on RQ2, really large P2M plants which could impact the sector intensely can be deployed at INPs (in Hungary). Results also showed that these large P2M plants with CC can have a better cost–benefit ratio than smaller P2M plants at ABPs or BEPs if CC costs would decrease significantly. If one considers that P2M at INPs are not only relevant by their size but by the commissioned number of them, and emitted CO_2 (energy supply and industry together was responsible for 48.3% of the greenhouse gas emission, agriculture for only 11.3% in Europe in 2014 [91]), CO_2 reuse with parallel energy storage of P2M at INPs can lead to disruption, but only if CC costs would radically fall. If INPs can be the high-end market for P2M, this is because of the better cost–benefit ratio, the higher potential of a single P2M plant size, and the higher number of possible plants (market potential). In contrast, WWTPs, ABPs, and BEPs representing the low-end segment of the market can be more suitable for P2M implementations in grid-scale. Nevertheless, the applicability of the revised theory about disruptive innovation (not technology) by Christensen et al. [38] is limited in this study, as incumbents (established large companies with sustaining technologies) who may overlook the low-end segment and will be challenged by disruption were not identified. Probably, this is because of the relatively new market generated by sustainability efforts.

Figure 5 summarizes these findings aligned with the research framework. The answers to the research questions are the following:

RQ1: What are the key attributes of P2M for potential technology adopters and how can they be evaluated compared to other (maybe sustaining) technologies? The key attributes of P2M are (1) producing renewable gas or another energy carrier different from electricity, (2) providing grid balancing services, (3) short-term and long-term energy storage, and (4) direct decarbonization. This attribute package is unique with a parallel function for decarbonization and energy storage.

RQ2: What is the largest P2M plant size possible at different types of sites and what sites are preferred for large-scale P2M deployments as possible low-end and high-end segments? Based on the empirical data collection and analysis, the largest possible P2M plant size was identified at an INP (30 MW$_{el}$), which would need CC solutions as well. Because of the larger P2M potential, INPs are the high-end segments for P2M; ABPs, BEPs, and WWTPs are the low-end segments (with lower P2M potential, but without CC).

RQ2: Which environmental factors and technological advancements could lead to superior performance compared to other (maybe sustaining) technologies and accelerate the process of P2M implementation? A significant decrease of CC costs could enable the disruption potential of the P2M technology in the future, along with further growth of renewable energy production, decarbonization incentives, and significant support of the regulatory environment (e.g., on the regulatory side, the volume of carbon taxes which can be as much important as CC costs).

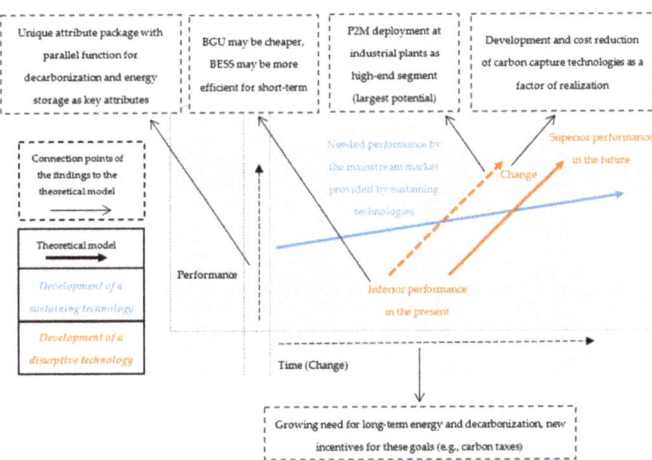

Figure 5. The disruption potential of P2M technology (main conclusions aligned with the research framework).

In summary, due to its unique attribute package, the P2M technology today is rather a value innovation [92], and a potentially disruptive technology of the future. Figure 6 shows the unique attribute package of P2M as a value curve indicating the value innovation. The identified unique attributes of P2M (the parallel CO$_2$ reuse and the energy storage potential) are in line with former micro-level achievements and energy evaluations as well. For example, Castellani et al. [93] found that methane production has a higher energy storage capability than methanol production, which can be the basis of the P2L process.

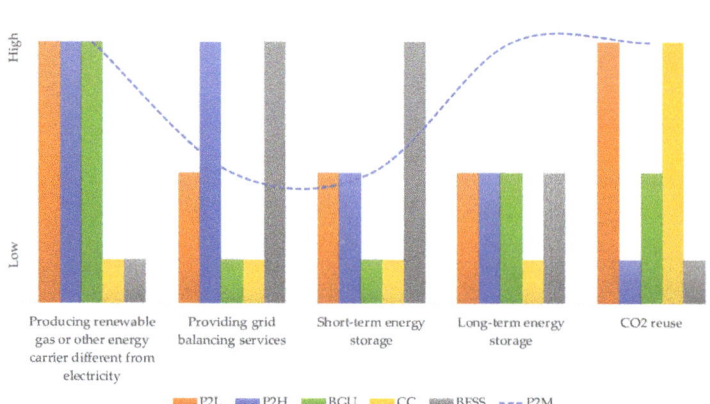

Figure 6. P2M attribute package as a new value curve (relative values).

P2M and CC together could become disruptive in the future as CC costs would decrease and volatile renewable energy production and decarbonization pressure would increase further. Regarding CC technologies, oxy-combustion is seen as a promising and cost-effective method in the literature [94], but regarding the oxygen by-product of the electrolysis in the P2M process, it could lead to even more synergies in theory, which could be re-used in the oxy-fuel CC process.

Finally, two other points should be highlighted based on the empirical results. First, in contrast to underlying assumptions, no "competition" between catalytic or biological methanation, nor between AEL or PEMEL, was relevant from a disruptive point of view. Second, findings suggest that P2H, P2M, and P2L, and even BESS, can be parts of an integrated energy system at a large industrial company, providing short-term and long-term energy storage, renewable energy production with market-flexibility (hydrogen, methane or hydrocarbon fuel), and capability for grid-balancing.

5. Conclusions

The starting point of this study was that P2M could be considered a disruptive technology because of its predicted future impact on the energy sector, and the new opportunities and new challenges it generates. The disruptiveness of the technology, however, hasn't been assessed yet. The working hypothesis of the study was that P2M could become disruptive. This can be accepted based on the results. Using hybrid (quantitative and qualitative) methodology, it was concluded that P2M currently is rather a value innovation due to its unique attribute package, the combined seasonal energy storage and direct decarbonization function. Besides that, it was demonstrated that P2M has the potential of becoming a disruptive technology if associated with CC technologies, and if the current CAPEX volumes related to this technology would decrease significantly. It was also presumed that renewable energy generation would continue to grow because the largest P2M potential can be identified at those industrial plants where CO_2 should be captured from flue gas. This conclusion has another practical contribution as well, by highlighting that CC technology developments should get a higher priority to completely exploit the disruption potential of the P2M technology. From a practical point of view, findings suggest that agricultural biogas plants and bioethanol plants with efficiently usable carbon sources, as well as industrial sites with carbon capture solutions, could be equally suitable from the aspect of CO_2 input for building the largest P2M plant worldwide, which could also be located in Hungary (over 6 MW_{el}). For seasonal energy storage, however, agricultural biogas plants and bioethanol plants are not as promising as formerly presented large wastewater treatment plants [13], while, in the case of industrial adopters, the costs of carbon capture decrease the economic attractiveness currently, which may change over time.

Nevertheless, some technological possibilities or alternatives might not have emerged because of the research framework. For example, while P2L was relevant for opening new market opportunities in the transportation sector, solutions for CNG or LNG production are one step ahead after P2M in the value chain. The combination of the P2M process and LNG production [95] would be, however, the competitor of P2L in the transforming transportation sector. Accordingly, future research could focus on the techno-economic comparisons of P2L and P2M + LNG. Another limitation of the study is that cost–benefit ratios were determined based on CAPEX, but hardly predictable operational expenses and revenue streams can accelerate or decelerate the possible disruption process of P2M. For example, the avoided carbon taxes could influence the unit costs of decarbonization at INPs, or the effects of other regulatory interventions (e.g., feed-in-tariffs, electricity prices) could be researched from this aspect, as well. Furthermore, the background of mainstream consumer needs could also be explored deeper. Finally, analyzing the synergies of oxy-combustion and P2H/P2M process could be a relevant topic in relation to increasing the market attractiveness of seasonal energy storage and decarbonization for multiple stakeholders, which this research also focused on.

Author Contributions: Conceptualization, G.P., Z.C., and M.Z.; methodology, M.Z.; validation, G.P. and Z.C., formal analysis, Z.C. and M.Z.; investigation, G.P., Z.C., and M.Z.; data curation, M.Z.; writing—original draft preparation, G.P. and M.Z.; writing—review and editing, Z.C.; visualization, M.Z.; supervision, Z.C.; project administration, Z.C. All authors have read and agreed to the published version of the manuscript.

Funding: This research received no external funding.

Acknowledgments: The authors would like to thank Hiventures Zrt./State Fund for Research and Development and Innovation for their investment that enabled this research.

Conflicts of Interest: The authors declare no conflict of interest.

Appendix A

kW_{el}	5000	12,500	15,000	30,000	Source
CAPEX estimation of large-scale P2M plants with biological methanation for 2025:					
2025					
Electrolyzer system (PEMEC) (thEUR/kW_{el})	0,90	0.85	0.80	0.75	[40]
Methanation system (biological) (thEUR/kW_{el})	0.35	0.30	0.25	0.2	[40]
Infrastructure, installation, storage for gas puffer (H_2, CO_2), injection (thEUR/kW_{el})	0.45	0.40	0.35	0.30	ca. 20% of CAPEX ([55], p. 34)
Project development, planning, expert services, quality management (+ %)	28%	28%	28%	28%	([55], p. 34)
2030					
Electrolyzer system (PEMEC) (thEUR/kW_{el})	0.90	0.85	0.80	0.75	[40]
Methanation system (biological) (thEUR/kW_{el})	0.35	0.30	0.25	0.2	[40]
Infrastructure, installation, storage for gas puffer (H2, CO_2), injection (thEUR/kW_{el})	0.45	0.40	0.35	0.30	ca. 20% of CAPEX ([55], p. 34]
Project development, planning, expert services, quality management (+ %)	28%	28%	28%	28%	([55], p. 34)

Appendix B

Detailed data of large-scale P2M plants with 8000 h/year operation:

	WWTP	ABP	BEP	INP
Size (MW$_{el}$)	5	12.5	15	30
Converted CO_2/year (tons)	4240	10,600	12,720	25,440
Produced CH_4/year (MWh)	21,815	54,538	65,445	130,890
Converted CO_2/ 20 years (tons)	84,800	212,000	254,400	508,800
Produced CH_4/ 20 years (MWh)	436,300	1,090,750	1,308,900	2,617,800
P2M CAPEX (EUR, prediction for 2025)	10,880,000	24,800 000	26,880,000	48,000,000
Cost of carbon capture (20 years) (EUR)	-	-	-	20,352,000
Unit cost of decarbonization (20 years) (EUR/t)	128	117	106	134
Unit cost of renewable gas production (20 years) (EUR/MWh)	25	23	21	26

Abbreviations

ABP	Agricultural biogas plant
AEL	Alkaline electrolysis
BEP	Bioethanol plant
BESS	Battery energy storage systems
BGU	Biogas upgrading
CAPEX	Capital expenditures
CC	Carbon capture
CHP	Combined heat and power (unit)
CNG	Compressed natural gas
EMG-BES	Bioelectrochemical system for electromethanogenesis
IGCC	Integrated gasification combined cycle
INP	Industrial plant
LIB	Lithium-ion battery
LNG	Liquified natural gas
NGCC	Natural gas combined cycle
P2G	Power-to-Gas
P2H	Power-to-Hydrogen
P2L	Power-to-Liquid
P2M	Power-to-Methane
PC	Pulverized coal
PEMEL	Polymer electrolyte membrane electrolysis
SOEL	Solid-oxide electrolysis
TRL	Technological readiness level
WWTP	Wastewater treatment plant

References

1. Lund, H.; Østergaard, P.A.; Connolly, D.; Ridjan, I.; Mathiesen, B.V.; Hvelplund, F.; Thellufsen, J.Z.; Sorknæs, P. Energy Storage and Smart Energy Systems. *Int. J. Sustain. Energy Plan. Manag.* **2016**, *11*, 3–14. [CrossRef]
2. Ergüden, E.; Çatlıoglu, E. Sustainability Reporting Practiceses In Energy Companies With Topsis Method. *J. Account. Financ.* **2016**, *71*, 201–221.
3. Bollino, C.A.; Madlener, R. Foreword to the Special Issue on High Shares of Renewable Energy Sources and Electricity Market Reform. *Energy J.* **2016**, *37*, 1–4. [CrossRef]
4. Adil, A.M.; Ko, Y. Socio-technical evolution of Decentralized Energy Systems: A critical review and implications for urban planning and policy. *Renew. Sustain. Energy Rev.* **2016**, *57*, 1025–1037. [CrossRef]
5. Alagoz, B.B.; Kaygusuz, A. Dynamic energy pricing by closed-loop fractional-order PI control system and energy balancing in smart grid energy markets. *Trans. Inst. Meas. Control* **2016**, *38*, 565–578. [CrossRef]
6. Costa-Campi, M.; Duch-Brown, N.; García-Quevedo, J. R & D drivers and obstacles to innovation in the energy industry. *Energy Econ.* **2014**, *46*, 20–30. [CrossRef]
7. European Commission. *Report From The Commission To The European Parliament, The Council, The European Economic And Social Committee And The Committee Of The Regions. Renewable Energy Progress Report*; European Commission: Brussels, Belgium, 2020.
8. European Commission. *European Climate Pact—Communication From The Commission To The European Parliament, The Council, The European Economic And Social Committee And The Committee Of The Regions*; European Commission: Brussels, Belgium, 2020.

9. Pintér, G. The Potential Role of Power-to-Gas Technology Connected to Photovoltaic Power Plants in the Visegrad Countries—A Case Study. *Energies* **2020**, *13*, 6408. [CrossRef]
10. Götz, M.; Lefebvre, J.; Mörs, F.; Koch, A.M.; Graf, F.; Bajohr, S.; Reimert, R.; Kolb, T. Renewable Power-to-Gas: A technological and economic review. *Renew. Energy* **2016**, *85*, 1371–1390. [CrossRef]
11. STORE&GO. The STORE&GO Demonstration Sites. Available online: https://www.storeandgo.info/demonstration-sites (accessed on 11 February 2021).
12. Bailera, M.; Lisbona, P.; Romeo, L.M.; Espatolero, S. Power to Gas projects review: Lab, pilot and demo plants for storing renewable energy and CO2. *Renew. Sustain. Energy Rev.* **2017**, *69*, 292–312. [CrossRef]
13. Guilera, J.; Morante, J.R.; Andreu, T. Economic viability of SNG production from power and CO_2. *Energy Convers. Manag.* **2018**, *162*, 218–224. [CrossRef]
14. Peters, R.; Baltruweit, M.; Grube, T.; Samsun, R.C.; Stolten, D. A techno economic analysis of the power to gas route. *J. CO2 Util.* **2019**, *34*, 616–634. [CrossRef]
15. European Commission. *Study on Energy Storage—Contribution to the Security of the Electricity Supply in Europe*; European Commission: Brussels, Belgium, 2020.
16. Csedő, Z.; Zavarkó, M. The role of inter-organizational innovation networks as change drivers in commercialization of disruptive technologies: The case of power-to-gas. *Int. J. Sustain. Energy Plan. Manag.* **2020**, *28*, 53–70. [CrossRef]
17. Csedő, Z.; Sinóros-Szabó, B.; Zavarkó, M. Seasonal Energy Storage Potential Assessment of WWTPs with Power-to-Methane Technology. *Energies* **2020**, *13*, 4973. [CrossRef]
18. Cantwell, J. Innovation and competitiveness. In *The Oxford Handbook of Innovation*; Fagerber, J., Mowery, D.C., Nelson, R.R., Eds.; Oxford University Press: New York, NY, USA, 2005; pp. 543–567.
19. Feurer, R.; Chaharbaghi, K. Defining Competitiveness: A Holistic Approach. *Manag. Decis.* **1994**, *32*, 49–58. [CrossRef]
20. Schwab, K. *The Global Competitiveness Report 2019*; World Economic Forum: Geneva, Switzerland, 2019.
21. Almutairi, J.; Aldossary, M. Modeling and Analyzing Offloading Strategies of IoT Applications over Edge Computing; Joint Clouds. *Symmetry* **2021**, *13*, 402. [CrossRef]
22. Diener, F.; Špaček, M. Digital Transformation in Banking: A Managerial Perspective on Barriers to Change. *Sustainability* **2021**, *13*, 2032. [CrossRef]
23. Lekan, A.; Clinton, A.; Owolabi, J. The Disruptive Adaptations of Construction 4.0 and Industry 4.0 as a Pathway to a Sustainable Innovation and Inclusive Industrial Technological Development. *Buildings* **2021**, *11*, 79. [CrossRef]
24. Servi, M.; Zulli, A.; Volpe, Y.; Furferi, R.; Puggelli, L.; Messineo, A.; Ghionzoli, M.; Facchini, F. Handheld Optical System for Pectus Excavatum Assessment. *Appl. Sci.* **2021**, *11*, 1726. [CrossRef]
25. Ullah, N.; Al-Rahmi, W.; Alzahrani, A.; Alfarraj, O.; Alblehai, F. Blockchain Technology Adoption in Smart Learning Environments. *Sustainability* **2021**, *13*, 1801. [CrossRef]
26. Santamaria, B.M.; Gonzalo, F.A.; Aguirregabiria, B.; Ramos, J.H. Evaluation of Thermal Comfort and Energy Consumption of Water Flow Glazing as a Radiant Heating and Cooling System: A Case Study of an Office Space. *Sustainability* **2020**, *12*, 7596. [CrossRef]
27. Radu, L.-D. Disruptive Technologies in Smart Cities: A Survey on Current Trends and Challenges. *Smart Cities* **2020**, *3*, 51. [CrossRef]
28. Yigitcanlar, T.; DeSouza, K.C.; Butler, L.; Roozkhosh, F. Contributions and Risks of Artificial Intelligence (AI) in Building Smarter Cities: Insights from a Systematic Review of the Literature. *Energies* **2020**, *13*, 1473. [CrossRef]
29. Enescu, F.M.; Bizon, N.; Onu, A.; Răboacă, M.S.; Thounthong, P.; Mazare, A.G.; Șerban, G. Implementing Blockchain Technology in Irrigation Systems That Integrate Photovoltaic Energy Generation Systems. *Sustainability* **2020**, *12*, 1540. [CrossRef]
30. Ullah, N.; Alnumay, W.S.; Al-Rahmi, W.M.; Alzahrani, A.I.; Al-Samarraie, H. Modeling Cost Saving and Innovativeness for Blockchain Technology Adoption by Energy Management. *Energies* **2020**, *13*, 4783. [CrossRef]
31. Zeng, Y.; Dong, P.; Shi, Y.; Li, Y. On the Disruptive Innovation Strategy of Renewable Energy Technology Diffusion: An Agent-Based Model. *Energies* **2018**, *11*, 3217. [CrossRef]
32. Müller, J.M.; Kunderer, R. Ex-Ante Prediction of Disruptive Innovation: The Case of Battery Technologies. *Sustainability* **2019**, *11*, 5229. [CrossRef]
33. Schiebahn, S.; Grube, T.; Robinius, M.; Tietze, V.; Kumar, B.; Stolten, D. Power to gas: Technological overview, systems analysis and economic assessment for a case study in Germany. *Int. J. Hydrog. Energy* **2015**, *40*, 4285–4294. [CrossRef]
34. Schäfer, M.; Gretzschel, O.; Steinmetz, H. The Possible Roles of Wastewater Treatment Plants in Sector Coupling. *Energies* **2020**, *13*, 2088. [CrossRef]
35. Breyer, C.; Tsupari, E.; Tikka, V.; Vainikka, P. Power-to-Gas as an Emerging Profitable Business Through Creating an Integrated Value Chain. *Energy Procedia* **2015**, *73*, 182–189. [CrossRef]
36. Bower, J.L.; Christensen, C.M. Disruptive technologies: Catching the wave. *Harvard Business Review*, 19 February 1995, 43–53.
37. Christensen, C.M.; Raynor, M.E.; McDonald, R. What is disruptive innovation? *Harvard Business Review*, 20 December 2015, 44–53.
38. Govindarajan, V.; Kopalle, P.K. Disruptiveness of innovations: Measurement and an assessment of reliability and validity. *Strateg. Manag. J.* **2006**, *27*, 189–199. [CrossRef]
39. Böhm, H.; Zauner, A.; Rosenfeld, D.C.; Tichler, R. Projecting cost development for future large-scale power-to-gas implementations by scaling effects. *Appl. Energy* **2020**, *264*, 114780. [CrossRef]

40. Ministry of Innovation and Technology of Hungary. *National Energy Strategy 2030*; Ministry of Innovation and Technology of Hungary: Budapest, Hungary, 2020.
41. Pintér, G.; Zsiborács, H.; Baranyai, N.H.; Vincze, A.; Birkner, Z. The Economic and Geographical Aspects of the Status of Small-Scale Photovoltaic Systems in Hungary—A Case Study. *Energies* **2020**, *13*, 3489. [CrossRef]
42. Lewin, K. Action research and minority problems. *J. Soc. Issues* **1946**, *2*, 34–46. [CrossRef]
43. Reason, P. *Handbook of Action Research: Participative Inquiry and Practice*; SAGE: London, UK, 2001.
44. McNiff, J. *Action Research—Principles and Practice*; Routledge: London, UK, 2013.
45. Mets, L. Methanobacter Thermoautotrophicus Strain and Variants. Thereof. Patent EP2661511B1, 5 January 2012.
46. Lee, D.-Y.; Mehran, M.T.; Kim, J.; Kim, S.; Lee, S.-B.; Song, R.-H.; Ko, E.-Y.; Hong, J.-E.; Huh, J.-Y.; Lim, T.-H. Scaling up syngas production with controllable H2/CO ratio in a highly efficient, compact, and durable solid oxide coelectrolysis cell unit-bundle. *Appl. Energy* **2020**, *257*, 114036. [CrossRef]
47. Wang, L.; Pérez-Fortes, M.; Madi, H.; Diethelm, S.; Van Herle, J.; Maréchal, F. Optimal design of solidoxide electrolyzer based power-to-methane systems: A comprehensive comparison between steam electrolysis and co-electrolysis. *Appl. Energy* **2018**, *211*, 1060–1079. [CrossRef]
48. Schmidt, P.; Weindorf, W. *Power-to-Liquids—Potentials and Perspectives for the Future Supply of Renewable Aviation Fuel*; German Environment Agency: Munich, Germany, 2016.
49. Lochbrunner, A. *PtG Demonstration Plant Solothurn Commissioned—Innovative Large-Scale Energy Storage Technologies and Power-to-Gas Concepts After Optimisation*; STORE&GO Project: Karlsruhe, Germany, 2018.
50. IRENA. *Hydrogen From Renewable Power: Technology Outlook for the Energy Transition*; IRENA: Abu Dhabi, UAE, 2018.
51. Buttler, A.; Spliethoff, H. Current status of water electrolysis for energy storage, grid balancing and sector coupling via power-to-gas and power-to-liquids: A review. *Renew. Sustain. Energy Rev.* **2018**, *82*, 440–2454. [CrossRef]
52. Abu Bakar, W.; Ali, R.; Toemen, S. Catalytic methanation reaction over supported nickel-ruthenium oxide base for purification of simulated natural gas. *Sci. Iran.* **2012**, *19*, 525–534. [CrossRef]
53. Jaffar, M.M.; Nahil, M.A.; Williams, P.T. Parametric Study of CO_2 Methanation for Synthetic Natural Gas Production. *Energy Technol.* **2019**, *7*, 1900795. [CrossRef]
54. Frontera, P.; Macario, A.; Ferraro, M.; Antonucci, P. Supported Catalysts for CO2 Methanation: A Review. *Catalysts* **2017**, *7*, 59. [CrossRef]
55. Leeuwen, C.; Zauner, A. *Report on the Costs Involved with P2G Technologies and Their Potentials Across the EU*; STORE&GO Project: Karlsruhe, Germany, 2018.
56. Electrochaea GmbH. How the Technology Works. 2019. Available online: http://www.electrochaea.com/technology (accessed on 18 March 2019).
57. Sinóros-Szabó, B.; Zavarkó, M.; Popp, F.; Grima, P.; Csedő, Z. Biomethane production monitoring and data analysis based on the practical operation experiences of an innovative power-to-gas benchscale prototype. *Acta Agrar. Debr.* **2018**, *150*, 399–410. [CrossRef]
58. Györke, G.; Groniewsky, A.; Imre, A.R. A Simple Method of Finding New Dry and Isentropic Working Fluids for Organic Rankine Cycle. *Energies* **2019**, *12*, 480. [CrossRef]
59. Ceballos-Escalera, A.; Molognoni, D.; Bosch-Jimenez, P.; Shahparasti, M.; Bouchakour, S.; Luna, A.; Guisasola, A.; Borràs, E.; Della Pirriera, M. Bioelectrochemical systems for energy storage: A scaled-up power-to-gas approach. *Appl. Energy* **2020**, *260*, 114138. [CrossRef]
60. Ács, N.; Szuhaj, M.; Wirth, R.; Bagi, Z.; Maróti, G.; Rákhely, G.; Kovács, K.L. Microbial Community Rearrangements in Power-to-Biomethane Reactors Employing Mesophilic Biogas Digestate. *Front. Energy Res.* **2019**, *7*, 1–15. [CrossRef]
61. Laude, A.; Ricci, O.; Bureau, G.; Royer-Adnot, J.; Fabbri, A. CO2 capture and storage from a bioethanol plant: Carbon and energy footprint and economic assessment. *Int. J. Greenh. Gas Control* **2011**, *5*, 1220–1231. [CrossRef]
62. Chauvy, R.; Dubois, L.; Lybaert, P.; Thomas, D.; De Weireld, G. Production of synthetic natural gas from industrial carbon dioxide. *Appl. Energy* **2020**, *260*, 114249. [CrossRef]
63. Sinóros-Szabó, B. Evaluation of Biogenic Carbon Dioxide Market and Synergy Potential for Commercial-Scale Power-to-Gas Facilities in Hungary. In Proceedings of the Innovációs Kihívások a XXI. Században, Innovation Challenges in the XXI. Century, LXI. Georgikon Days Conference, Keszthely, Hungary, 3–4 October 2019; Pintér, G., Csányi, S., Zsiborács, H., Eds.; Georgikon Faculty—University of Pannonia: Keszthely, Hungary, 2019; pp. 371–380, ISBN 9789633961308.
64. Strauss, A.L.; Corbin, J. *Basics of Qualitative Research: Techniques and Procedures for Developing Grounded Theory*; Sage Publications: Thousand Oaks, CA, USA, 1998.
65. Danneels, E. Trying to become a different type of company: Dynamic capability at Smith Corona. *Strateg. Manag. J.* **2010**, *32*, 1–31. [CrossRef]
66. Bingham, C.B.; Heimeriks, K.H.; Schijven, M.; Gates, S. Concurrent learning: How firms develop multiple dynamic capabilities in parallel. *Strateg. Manag. J.* **2015**, *36*, 1802–1825. [CrossRef]
67. Tripsas, M.; Gavetti, G. Capabilities, cognition, and inertia: Evidence from digital imaging. *Strateg. Manag. J.* **2000**, *21*, 1147–1161. [CrossRef]
68. Angelidaki, I.; Treu, L.; Tsapekos, P.; Luo, G.; Campanaro, S.; Wenzel, H.; Kougias, P.G. Biogas upgrading and utilization: Current status and perspectives. *Biotechnol. Adv.* **2018**, *36*, 452–466. [CrossRef] [PubMed]

69. Gabnai, Z. Energy alternatives in large-scale wastewater treatment. *Appl. Stud. Agribus. Commer.* **2017**, *11*, 141–146. [CrossRef]
70. van Melle, T.; Peters, D.; Cherkasky, J.; Wessels, R.; Mir, G.U.R.; Hofsteenge, W. *Gas for Climate—How Gas Can Help to Achieve the Paris Agreement Target in an Affordable Way*; Ecofys: Utrecht, The Netherlands, 2018.
71. Varone, A.; Ferrari, M. Power to liquid and power to gas: An option for the German Energiewende. *Renew. Sustain. Energy Rev.* **2015**, *45*, 207–2018. [CrossRef]
72. Sunfire. Norsk E-Fuel Is Planning Europe's First Commercial Plant For Hydrogen-Based Renewable Aviation Fuel In Norway. 2020. Available online: https://www.sunfire.de/en/news/detail/norsk-e-fuel-is-planning-europes-first-commercial-plant-for-hydrogen-based-renewable-aviation-fuel-in-norway (accessed on 12 February 2021).
73. Yong, P. Theoretical discovery of high capacity hydrogen storage metal tetrahydrides. *Int. J. Hydrog. Energy* **2019**, *44*, 8153–18158. [CrossRef]
74. Yang, Y.; Bremner, S.; Menictas, C.; Kay, M. Battery energy storage system size determination in renewable energy systems: A review. *Renew. Sustain. Energy Rev.* **2018**, *91*, 109–125. [CrossRef]
75. Faessler, B.; Schuler, M.; Preißinger, M.; Kepplinger, P. Battery Storage Systems as Grid-Balancing Measure in Low-Voltage Distribution Grids with Distributed Generation. *Energies* **2017**, *10*, 2161. [CrossRef]
76. Gibbins, J.; Chalmers, H. Carbon capture and storage. *Energy Policy* **2008**, *36*, 4317–4322. [CrossRef]
77. Leung, D.Y.C.; Caramanna, G.; Maroto-Valer, M.M. An overview of current status of carbon dioxide capture and storage technologies. *Renew. Sustain. Energy Rev.* **2014**, *39*, 426–443. [CrossRef]
78. Zhang, X.; Bauer, C.; Mutel, C.L.; Volkart, K. Life Cycle Assessment of Power-to-Gas: Approaches, system variations and their environmental implications. *Appl. Energy* **2017**, *190*, 326–338. [CrossRef]
79. Fan, J.-L.; Xu, M.; Li, F.; Yang, L.; Zhang, X. Carbon capture and storage (CCS) retrofit potential of coal-fired power plants in China: The technology lock-in and cost optimization perspective. *Appl. Energy* **2018**, *229*, 326–334. [CrossRef]
80. Wilberforce, T.; Olabi, A.; Sayed, E.T.; Elsaid, K.; Abdelkareem, M.A. Progress in carbon capture technologies. *Sci. Total Environ.* **2021**, *761*, 143203. [CrossRef] [PubMed]
81. Wenge, C.; Pietracho, R.; Balischewski, S.; Arendarski, P.; Lombardi, P.; Komarnicki, P.; Kasprzyk, L. Multi Usage Applications of Li-Ion Battery Storage in a Large Photovoltaic Plant: A Practical Experience. *Energies* **2020**, *13*, 4590. [CrossRef]
82. Sitompul, S.; Hanawa, Y.; Bupphaves, V.; Fujita, G. State of Charge Control Integrated with Load Frequency Control for BESS in Islanded Microgrid. *Energies* **2020**, *13*, 4657. [CrossRef]
83. Bienert, K.; Schumacher, B.; Arboleda, M.R.; Billig, E.; Shakya, S.; Rogstrand, G.; Zieliński, M.; Dębowski, M. Multi-Indicator Assessment of Innovative Small-Scale Biomethane Technologies in Europe. *Energies* **2019**, *12*, 1321. [CrossRef]
84. Nguyen, T.P.; Kim, I.T. Ag Nanoparticle-Decorated MoS2 Nanosheets for Enhancing Electrochemical Performance in Lithium Storage. *Nanomaterials* **2021**, *11*, 626. [CrossRef]
85. Drünert, S.; Neuling, U.; Zitscher, T.; Kaltschmitt, M. Power-to-Liquid fuels for aviation—Processes, resources and supply potential under German conditions. *Appl. Energy* **2020**, *277*, 115578. [CrossRef]
86. Hu, G.; Chen, C.; Lu, H.T.; Wu, Y.; Liu, C.; Tao, L.; Men, Y.; He, G.; Li, K.G. A Review of Technical Advances, Barriers, and Solutions in the Power to Hydrogen (P2H) Roadmap. *Engineering* **2020**, *6*, 1364–1380. [CrossRef]
87. Kapetaki, Z.; Miranda Barbosa, E. *Carbon Capture Utilisation and Storage Technology Development Report 2018*; European Commission: Luxembourg, 2019; ISBN 978-92-76-12440-5. [CrossRef]
88. Bauen, A.; Bitossi, N.; German, L.; Harris, A.; Leow, K. Sustainable Aviation Fuels: Status, challenges and prospects of drop-in liquid fuels, hydrogen and electrification in aviation. *Johns. Matthey Technol. Rev.* **2020**, *64*, 263–278. [CrossRef]
89. Khan, I.U.; Othman, M.H.D.; Hashim, H.; Matsuura, T.; Ismail, A.; Rezaei-DashtArzhandi, M.; Azelee, I.W. Biogas as a renewable energy fuel—A review of biogas upgrading, utilisation and storage. *Energy Convers. Manag.* **2017**, *150*, 277–294. [CrossRef]
90. Kucevic, D.; Tepe, B.; Englberger, S.; Parlikar, A.; Mühlbauer, M.; Bohlen, O.; Jossen, A.; Hesse, H. Standard battery energy storage system profiles: Analysis of various applications for stationary energy storage systems using a holistic simulation framework. *J. Energy Storage* **2020**, *28*, 101077. [CrossRef]
91. European Environmental Agency. Sectoral Greenhouse Gas Emissions by IPCC Sector. 2016. Available online: https://www.eea.europa.eu/data-and-maps/daviz/change-of-co2-eq-emissions-2#tab-dashboard-01 (accessed on 3 March 2021).
92. Kim, W.C.; Mauborgne, R. The Strategic logic of high growth. *Harvard Business Review* 103, 19 February 1997, p. 112.
93. Castellani, B.; Gambelli, A.M.; Morini, E.; Nastasi, B.; Presciutti, A.; Filipponi, M.; Nicolini, A.; Rossi, F. Experimental Investigation on CO2 Methanation Process for Solar Energy Storage Compared to CO_2-Based Methanol Synthesis. *Energies* **2017**, *10*, 855. [CrossRef]
94. Wu, F.; Argyle, M.D.; Dellenback, P.A.; Fan, M. Progress in O2 separation for oxy-fuel combustion—A promising way for cost-effective CO_2 capture: A review. *Prog. Energy Combust. Sci.* **2018**, *67*, 188–205. [CrossRef]
95. Imre, A.R.; Kustán, R.; Groniewsky, A. Thermodynamic Selection of the Optimal Working Fluid for Organic Rankine Cycles. *Energies* **2019**, *12*, 2028. [CrossRef]

Article

Seasonal and Multi-Seasonal Energy Storage by Power-to-Methane Technology

Kristóf Kummer [1] and Attila R. Imre [1,2,*]

[1] Department of Energy Engineering, Faculty of Mechanical Engineering, Budapest University of Technology and Economics, Műegyetem rkp. 3, H-1111 Budapest, Hungary; kristof.kummer97@gmail.com
[2] Department of Thermohydraulics, Centre for Energy Research, POB. 49, H-1525 Budapest, Hungary
* Correspondence: imreattila@energia.bme.hu

Abstract: The time-range of applicability of various energy-storage technologies are limited by self-discharge and other inevitable losses. While batteries and hydrogen are useful for storage in a time-span ranging from hours to several days or even weeks, for seasonal or multi-seasonal storage, only some traditional and quite costly methods can be used (like pumped-storage plants, Compressed Air Energy Storage or energy tower). In this paper, we aim to show that while the efficiency of energy recovery of Power-to-Methane technology is lower than for several other methods, due to the low self-discharge and negligible standby losses, it can be a suitable and cost-effective solution for seasonal and multi-seasonal energy storage.

Keywords: Power-to-Gas; Power-to-Fuel; P2M; P2G; P2F; biomethanization

Citation: Kummer, K.; Imre, A.R. Seasonal and Multi-Seasonal Energy Storage by Power-to-Methane Technology. *Energies* 2021, *14*, 3265. https://doi.org/10.3390/en14113265

Academic Editor: Shusheng Pang

Received: 4 May 2021
Accepted: 28 May 2021
Published: 2 June 2021

Publisher's Note: MDPI stays neutral with regard to jurisdictional claims in published maps and institutional affiliations.

Copyright: © 2021 by the authors. Licensee MDPI, Basel, Switzerland. This article is an open access article distributed under the terms and conditions of the Creative Commons Attribution (CC BY) license (https://creativecommons.org/licenses/by/4.0/).

1. Short and Long Time Energy Storage

The purpose of energy storage is to store unused electricity for later use. The use can be done by recovering the available part of the stored electricity and using it. However, due to legislative changes, when the intermediate product (e.g., hydrogen) of the storage process is a fuel, it is also considered energy storage [1]. In the current article, we only consider the variant where both input and output become electricity; the possibility of using it as a fuel is only mentioned as an extra option where relevant.

Storage can be achieved in many different ways [2,3]; the simplest would perhaps be to store the electricity as electricity without modification (in supercapacitors or superconducting rings), but these solutions are generally expensive and have relatively small storage capacities.

Fortunately, there are other solutions with lower cost and/or larger storage capacity, but these methods require the electricity to be converted into another form of energy and then converted back. This back-and-forth conversion is costly and requires special equipment or facilities. One of these methods is mechanical energy storage, where the stored electricity is converted into either potential (e.g., pumped storage reservoirs) or kinetic (e.g., flywheel reservoirs) energy, and then this potential or kinetic energy is converted back into electricity using generators. Energy can also be stored chemically, using the initial electricity to produce a fuel or to increase the energy content of an existing fuel. Perhaps the best-known form of this group is the production of hydrogen by electrolysis of water, where the hydrogen can be stored and then used to recover electricity later in time, e.g., using fuel cells. For historical and technological reasons, electrochemical storage is a separate category, where reversible electrochemical processes are used to store and recover the energy; this is how rechargeable batteries work. We should also mention the so-called heat accumulators; heat accumulation is not usually classified as energy storage because usually neither the input nor the output "product" is electricity. Nowadays, this is changing. Sometimes there is so much excess electricity production, it is worth using it to produce heat and using it later (the input is then electricity). It is then possible—albeit with low efficiency—to produce electricity from the heat again later, e.g., by using the heat for the input of an Organic Rankine cycle [4].

Energy storage is mainly needed to compensate for the difference between fluctuating energy production (mostly caused by the changing weather condition) and fluctuating demand. As shown by Hiesl et al. in EUROSTAT data [5], the percentage of renewable-based electricity (excluding conventional hydropower) in the EU-28 has increased from 1% to 20%. In relative terms, the largest increase over the period was for solar (PV) generation. For these renewables (biomass, biogas, bio-liquid and other bio-derived waste, wind (off- and on-shore type), tidal, geothermal) and for the conventional, i.e., river-based—hydropower (not included in the survey), the weather dependence can be clearly observed. This dependence can lead to large variations in production even in the short term for solar and wind, but in other cases, a longer-term dependence can also be observed. For example, in the case of biological materials, the production of raw materials (quantity as well as quality) depends on the weather on a seasonal basis, while in the case of conventional hydropower, production is also affected by the weather (rainfall, drought) over a period of seven to ten months or seasonally. Surprisingly, even geothermal electricity generation is weather-dependent. For example, in ORC-based power plants, which are often used on these heat sources, the condenser temperature and the efficiency of the whole power plant are affected by weather-dependent variations in air or surface water [6].

In relation to storage or balancing problems, due to weather dependency, we tend to think of problems and solutions related to sub-hourly basis (e.g., clouds before the sun), daily basis (solar panels do not produce at night) or weekly basis (the drop in industrial consumption on Saturday-Sunday). For such storage tasks (both in terms of duration and capacity), battery-based systems such as Li-ion can be used. However, these types of storage are not suitable where seasonal (due to winter-summer production and consumption differences) or possibly longer-term (several years) storage is required, i.e., the task is actual storage, not the regulation of current fluctuations. One reason is their self-discharge, which causes the energy stored in them to decrease continuously and another is the very high storage capacity requirements that occur when storing on a seasonal or annual basis.

The discharge time (very often mislabeled as storage time, but storage can happen both in loaded or unloaded conditions) is often used to indicate how long the currently marketed types of a given storage method would be able to continuously supply the connected consumer, such a diagram is shown in Figure 1.

Using this kind of diagram, one can decide that what would be the available storage solutions to provide the average need for a given consumer (or group of consumers) for a given length of time, under normal discharging conditions. For example, for a consumer who needs 100 GWh electricity to cover its expected consumption for a month, pumped hydro- or Power-to-Hydrogen methods would be viable solutions, being the (100 GWh; 1 month) point in the common part of the green and blue region.

For discharge time—storage capacity diagrams, the output power is usually not defined. Although, it makes a difference whether a storage system has to supply a small residential building or an entire industrial estate. It is usually assumed that the maximum power or close to the maximum power of the already existing storage systems; the uncertainty in this is well hidden by the double logarithmic nature of the diagram. As an example, a commercially available 21 tons, container-sized sodium sulphur (NaS) battery unit has a maximum storage capacity of 1.2 MWh. The maximal charge/discharge power is 200 kW, but occasionally only half of this power is used [7]; thus, the discharge time is 6–12 h. So this type would be a small "blob" with a not sharp boundary between 1–1.2 MWh and 6–12 h within the grey ellipse in a discharge time vs. storage capacity diagram (Figure 1). From this figure, it is possible to determine how long a fully charged storage can supply the consumer from the start of discharge, assuming a more or less constant (or, because of the logarithmic scale, at least one order of magnitude) power output.

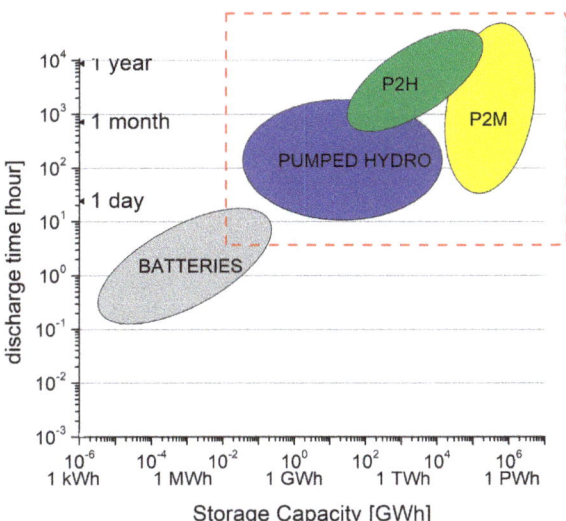

Figure 1. Schematic discharge time vs. storage capacity diagram for various storage methods, including batteries, pumped hydro, Power-to-Hydrogen and Power-to-Methane [5]. Long-term storage solutions are located inside the dashed rectangle.

Another time-related descriptor used for energy storage is the lifetime of the equipment itself. This is often given in terms of a maximum number of cycles (a cycle is a charge and discharge), in which case the lifetime can be obtained by multiplying this number of cycles by the average charge-discharge time. Another lifetime is the so-called shelf-life [8], showing the deterioration of a storage device from brand new to unusable, holding it in unused (and usually discharged or just partly charged) condition. The shelf-life is mainly given for batteries; many people are interested in how long an unused battery can be used, but less so in how long the dry bed of an unused pumped storage reservoir remains impermeable.

In this article, we would like to introduce a novel time-dependent quantity, which is not only time-dependent but also storage efficiency-dependent. This quantity shows that by filling a given type of storage and then storing it for t time after filling (without deliberate discharge, i.e., allowing only self-discharge), we get back a fraction of the energy stored as a function of time. This quantity will be important for seasonal, annual or multiannual storage, as it is not always the case that the photovoltaic energy, produced during a hot summer, can be recovered after 3–4 months (i.e., within a season) of unused storage.

In this paper, it will be proved that among the large-capacity storage methods, if the storage period exceeds half a year to one year, the so-called Power-to-Methane technology (in which methane is produced from water and carbon dioxide using stored electricity and then used to generate electricity at the time of storage) currently appears to be the most promising, from energetic and probably an economic point of view.

2. The Actual Discharge State Function

In this section, we introduce a novel quantity to help us to describe the actual state (the recoverable energy) for a given energy storage system. To understand the role of this new quantity, we need to generalize the term "self-discharge", which is mainly used for supercapacitor or battery storage. In self-discharge, the amount of energy stored in a storage device decreases even when it is unloaded; this usually happens in batteries due to a particular chemical reaction. For most battery types, this is a few tenths of a percent per day, but in some cases (such as in the case of a switched on redox liquid flow battery), it can be as much as 10% per day.

The generalization can be done in two different ways. First, in some cases, the so-called standby energy losses, which characterize the consumption of auxiliary equipment necessary for the operation of the storage, cannot be physically separated or should not be separated from the self-discharge losses; see, for example, the case of a sodium-sulphur battery. In this type, the dissipation heat of the self-discharge processes keeps the sodium and sulphur electrodes liquid during the 6–12 h charge-discharge cycles. While in a case where neither charge nor discharge occurs, this has to be done by an auxiliary heater, causing a loss of about 3% per day. The two types of losses can be physically separated, but since the effects of the two losses are the same, the separation does not make sense.

The second way to generalize the concept of self-discharge is the extension from capacitors and batteries to other storage devices. It is easy to see that evaporation and leakage losses in a pumped hydro storage, leakage of gas in a power-to-gas storage, leakage of the liquid or the degradation of the usually complex molecular structure in a power-to-liquid storage will cause losses similar to self-discharge of batteries, which are also time-dependent. Such losses can occur even in weight storage, although in the short term they may be due to a more random process (e.g., a few stones falling off a railway wagon used as weight storage), but over extremely long storing times, they may be of a more general nature (e.g., concrete elements of an abandoned weight tower start to crumble and erode).

The loss accumulates over time and is therefore given in units of percentage or part normalized to time (e.g., %/day), but this is only possible if the loss is stationary in time. When the speed of loss is not constant, it would be more appropriate to use a self-discharge function. If the strictly time-dependent self-discharge and other losses are summed, a time-dependent total storage loss can be obtained. Subtracting this from the amount of energy stored gives the energy that can be recovered from the storage. In this way, one can obtain an already time-dependent storage efficiency:

$$\frac{E_{ini} - (E_{sd}(t) + E_{sb}(t))}{E_{ini}} = \eta_s(t) \quad (1)$$

where $E_{sd}(t)$ is the time-dependent self-discharge function, $E_{sb}(t)$ is the time-dependent standby loss function, E_{ini} is the time-independent stored energy (at $t = 0$), and $\eta_s(t)$ is the now time-dependent storage efficiency including all losses and the discharging efficiency; this is what we call the Actual Discharge State Function or *ADSF*, which is a time-dependent function, correctly marked as *ADSF(t)*.

Now, the recovered energy (i.e., the amount recovered after full discharge) is

$$E_{ini} * \eta_s(t) = E_{ini} * ADSF(t) = E_d(t) \quad (2)$$

where $E_d(t)$ (subscript d stand for discharge) is also turns into a time-dependent quantity.

If the same amount of energy (for example, one "unit") is stored in two different types of storage devices, the *ADSF(t)* function gives the fraction of this energy that can be recovered if the discharge is started only t time after fully charging them; the two devices were unloaded during this t time, and the stored energy was reduced only by the generalized self-discharges. By comparing the *ADSF(t)* functions of these two storage facilities, it is easy to see which one will recover more energy later, starting the full discharge at any given time. This is demonstrated in Figure 2.

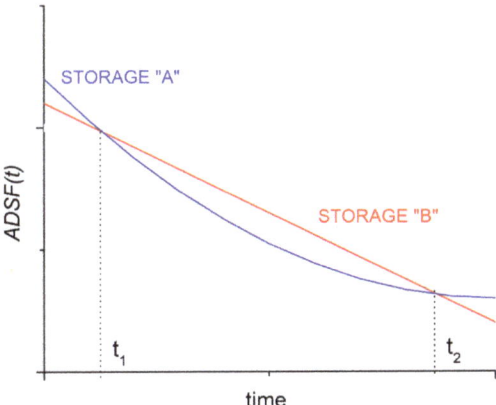

Figure 2. Using the time-dependent function $ADSF(t)$, it can be decided which kind of energy storage device is preferable when the discharging of the fully charged storage devices is started after an unloaded period. Starting discharge before time t_1, then storage A is better, starting discharge between t_1 and t_2, storage B is preferable while starting the discharge after t_2, storage A would again recover more electricity.

Obviously, besides the $ADSF(t)$ function, there are other quantities to be considered by choosing the proper storage technology, such as:
- installation and operating costs
- environmental and social criteria (pollution, social acceptance, etc.)
- power density
- energy density, total energy storage capacity.

Only the latter two are considered here. When comparing the $ADSF(t)$ function of two or more storage facilities, it will be indicated separately if one of them require an extremely large storage size to store the same amount of energy (energy density) or if it is not possible to build a storage size larger than a certain size for physical, economic or other reasons (e.g., the rarity of vanadium would make it difficult to build extra large vanadium redox storage facilities).

The $ADSF(t)$ function presented here is somewhat similar to the shelf-life, which is a time-independent but time-dimensional value given by manufacturers for batteries, referring to how long the storage device is functional when stored in an unloaded state. This quantity should also be time-dependent since it is possible that after six months, for example, the battery's storage capacity is 80% of the original capacity, while after 12 months, it is 60%. How long the storage device is considered to be usable also depends on the use; in some circumstances, 80% is not worth it for the user (in that case, the shelf-life of less than 6 months), in others cases, 60% is more than enough (in this case, shelf-life would be more than 12 months). The secondary use of batteries of electric cars is a good example, where after a while, they no longer fit for their original purpose but are still suitable for other purposes. Therefore, the original time-independent shelf-life (t_{sl}) can be generalized to obtain a time and remaining storage efficiency-dependent new shelf-life, where the latter "variable" could be a given limit value rather than a real variable. For example, the data pair of (t_{sl}^{60} = 1 year; t_{sl}^{20} = 2 years) that a given storage device would still be able to work on 60% of its original storage capacity after 1 year, and only 20% after two years).

An important distinction is that while the shelf-life is a quantity related to the unloaded storage facility (and this is also true for the time-dependent version), the $ADSF(t)$ function refers to the stored energy (also in unloaded state), which of course is also affected by the storage facility. Therefore, one can say that it is something, like the shelf-life of the stored energy.

Concerning the ADSF(t) function, for a given storage method, it can consist of several different time-dependent and time-independent parts. For example, in a pumped reservoir, the "self-discharge" itself introduces such terms; the evaporation loss depends on the external temperature and wind (this is time-dependent) and the current free surface area of the reservoir (this may be constant, but in dam reservoirs, it usually decreases as the volume in the reservoir decreases), while the seepage loss depends on the volume of water in the reservoir (the head of the water column, i.e., the pressure). Such a complex function is difficult to model, so for our comparison, we use a simplified (linear) ADSF(t) function. In this case, the time-dependent storage efficiency ($\eta_s(t)$) defined in Equation (1) will have a time-independent term (η_s) and a linear form of time-dependence. Therefore, the linearized ADSF(t) function takes the following form:

$$ADSF(t) = \eta_b(1 - \eta_s * t) \tag{3}$$

where, just like before, η_b is the efficiency of the conversion of stored energy back into electricity (i.e., the round-trip efficiency) and t is the time. In such a case, the curves in Figure 2 would become linear, and there would be only one intersection for two storage facilities, giving the time over which one reservoir is better for shorter storage and the other for longer storage. In this form, it can be seen that if we start discharging immediately after recharging (e.g., if we want to smooth PV output due to solar irradiance irregularities with a Li-ion battery), the ADSF(t = 0) is equal to the efficiency of converting the stored energy back to electricity, and then decreases linearly from there.

It can be seen that the actual ADSF(t) value for a given time can be increased in two ways; either by increasing the efficiency of the conversion efficiency upon discharge (e.g., in the Power-to-Methane case, by recovering the waste heat from the gas engine performing the conversion back in an ORC [9,10]), or by slowing the decrease, by reducing self-discharge (e.g., by better, more leakage-free storage of hydrogen in the case of Power-to-Hydrogen) or by reducing standby losses, such as in liquid electrode batteries by reducing heat loss through better insulation.

In the next section, some storage technologies are going to be presented by comparing their simplified (linear) ADSF(t) function to select which methods perform better than others for longer storage times. Then, based on the above two secondary criteria (energy density, total energy storage capacity), we will show which is the time interval of our interest (seasonal to multi-annual), the Power-to-Methane storage is likely to be the most appropriate.

3. Comparison of Various Energy Storage Methods

In this paper, we compare a few of the more well-known battery types, two Power-to-Gas storage types and one weight storage type. The traditional method for seasonal storage, pumped storage, is not considered here. On the one hand, its installation requires special natural conditions (i.e., it cannot be installed anywhere) [11], and on the other hand, there are countries (like Hungary), where installation of such kind of devices are strongly opposed for historical-political reasons [12].

Since the main objective is to place Power-to-Methane storage in the storage chain, the other types are only briefly described.

3.1. Batteries

The ADSF(t) functions of the following battery types will be discussed in this section:
- Lead-acid battery
- Nickel-metal hydride battery
- Lithium-ion (LiNMC/LiFePO$_4$) battery (new as well as second-life)
- Vanadium redox flow battery (in standby mode with flowing electrolyte and in offline mode with disconnected storage tanks)
- Sodium-sulphur (NaS) battery

We do not describe the first three types here in detail; all three types are well known, frequently used, and their characteristics can be found in the literature [3]. The values relevant for the estimation of the linearized $ADSF(t)$ functions are given in Table 1.

Table 1. Constants of the simplified (linear) $ADSF(t)$ function (Equation (3)). The values shown are for the best commercially available models for the type; some manufacturers' products may perform better or worse than this. Limits for these values are shown in Supplementary Materials Table S1.

Method	η_b	η_s/Day	Shelf-Life (Year)
Lead-acid battery	0.85	0.003	3–15
Nickel-metal hydride battery	0.80	0.005	5–10
Lithium-ion (LiNMC/LiFePO$_4$) battery	0.95	0.001	2–3
"Second-life" Lithium-ion battery	0.6	0.005	3–6
VRFB (offline)	0.75	0.2	20–30
VRFB (standby)	0.75	0	20–30
Sodium-Sulphur battery	0.85	0.068	15–25
Power-to-Hydrogen (with high-pressure gas storage)	0.75	0.01	>50
Power-to-Hydrogen (with cryogenic liquid storage)	0.75	0.006	>50
Power-to-Methane	0.33–0.5	0.000023	>50
Gravity storage	0.9	0.000064	>1000

Concerning Li-ion battery; this type is mostly used when high energy- and power-densities are needed; therefore bigger capacity Li-ion batteries are used mostly in transportation. For utility-scale seasonal storage, they would be "too good"; therefore, for this purpose, we are considering "second-life" batteries. These are batteries too much deteriorated for their original use, but still applicable for other purposes [13].

In the vanadium redox flow battery (VRFB), the chemical reaction takes place in a space, separated by a membrane (see Figure 3). Vanadium ions are present in the electrolyte in concentrations of a few mol/L, and the electrochemical reactions happen between different chemical valence states (V^{2+}/V^{3+} or V^{5+}/V^{4+}). The two types of electrolyte are stored in two separate tanks and can only come into contact with each other in the reaction space separated by a membrane. In practice, this type of battery is a small chemical factory; when the "intermediate products" are not required, the two types of electrolyte are stored without degradation, leakage or evaporation (i.e., self-discharge) in tanks, of which there may be more than one, and they may even be separated from the central, power-generating unit (i.e., as if they were liquid fuels in separate tanks). In this case, the battery is in a disconnected, offline state (the electrolyte is not circulated), with virtually zero self-discharge (until the plastic tanks break down and the electrolyte drains away). However, if it is flowing (i.e., it is in standby mode, ready for discharge), the daily self-discharge can be as high as 20%. The efficiency of the recovery is between 75–80%, including standby losses (e.g., pump operation in this case).

The sodium-sulphur (NaS) battery is a high-temperature, molten electrolyte battery; while the two electrodes (sodium and sulphur) are in a liquid, i.e., molten, state, the electrolyte is solid [7,14,15]. The internal temperature of the battery is at least 300 °C to keep the electrodes in liquid state. The battery belongs to the so-called energy batteries. Whereas, in power batteries (such as Li-ion batteries) the energy is delivered quickly (i.e., at high power), and for this type, the power is lower, but the total amount of energy stored is high. They are commercially available in container size; those made by NGK Insulators Ltd. of Japan can store 1.2 MWh and deliver this in six hours (or more) at a maximum power of 200 kW. The high temperature is provided by the dissipation heat generated by the self-discharge during continuous charge-discharge cycles. The overall conversion efficiency can in principle, reach 85%. In the unloaded state, one has to face a standby loss due to the necessary heating provided from the stored energy is 3.4 kW, i.e., 81 kWh per day, or 6.8% [16].

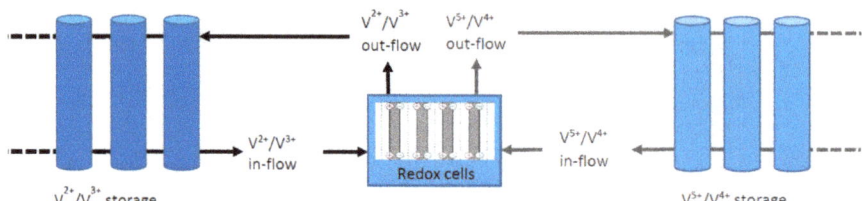

Figure 3. Schematic diagram of a Vanadium Redox Flow Battery with extendable and disconnectable electrolyte storage tanks.

3.2. Power-to-Gas Type Storage Systems

In Power-to-Fuel storage [3], electricity is used to produce a new fuel or convert an existing fuel to another with higher energy content. We are dealing with two sub-types within the method, both of them belonging to the Power-to-Gas group (i.e., the fuel produced is gaseous); one is hydrogen (Power-to-Hydrogen, P2H), and the other is methane (Power-to-Methane, P2M). The two methods are very closely related; in both cases, hydrogen is produced in the first step by hydrolysis using surplus electricity (to be stored). In the P2H method, that hydrogen is later used to generate electricity or as a vehicle fuel (but we are only looking at the electricity-storage-electricity type of methods). In pure form, it can be stored as a high-pressure gas or cryogenic liquid until reuse; alternatively, it can be stored in chemically bonded form (e.g., as ammonia) or mixed with natural gas [17,18]. In this method, the loss of hydrogen is responsible for the "self-discharge"; to estimate this value, we are considering high-pressure gas storage and cryogenic liquid storage separately.

In the P2M method, the hydrogen (produced by electrolysis using the surplus energy) and carbon dioxide (used from various sources) are used to produce methane by chemical [19] or biochemical [20] means; after the storage, the methane is used to generate electricity or as a vehicle fuel. In the present article, the biochemical version, which is less energy-intensive and therefore more efficient, is considered. It also has the advantage of being suitable for enriching methane-carbon dioxide mixtures (biogas, landfill gas) because, due to the low temperature, it can preserve the methane already present in the input gas. For conversion back to electricity, we are estimating a methane-to-electricity method of about 60% efficiency (e.g., an improved gas turbine), which gives a total storage efficiency of about 33%. It is also possible to convert the waste heat of electrolysis and methanization (approximately 30% of the incoming energy are lost in these two steps, part of these losses happens in the form of 60–70 °C waste heat) back into electricity by a low-temperature ORC process [21] and fed back into the electrolyzer, reducing the amount of energy input and thus increasing the storage efficiency with 1–2%. Also, it is possible to utilize the waste heat produced upon the recovery of the electricity, using a second ORC equipment. In this way, one might assume an upper limit for overall storage efficiency around 50%; we are discussing the two cases (33% and 50%) separately. In both cases, the methane would be stored in the natural gas network; the self-discharge, is thus, leakage from the network, the value of which was estimated from other data [22,23]. The steps of the whole cycle can be seen in Figure 4.

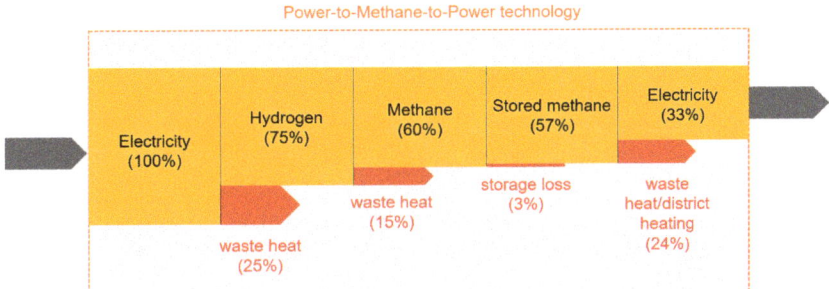

Figure 4. The steps of the Power-to-Methane-to-Power cycle; efficiencies and dissipative losses are marked.

3.3. Comparison of the Various ADSF(t) Functions

Relevant quantities to estimate the linearized $ADSF(t)$ functions for the storage technologies discussed here are listed in Table 1.

Based on these data, the $ADSF(t)$ vs. time function (Figure 5) can be plotted, showing what percentage of the initially stored electricity can be recovered (also as electricity) when the fully charged storage is discharged after an unloaded condition of t time. Because efficiencies can be different for various products within one type, thus, on Figure 5a,b, best and worst scenarios are marked for methods, where the efficiencies are moving in a wide range. Concerning these scenarios, we are considering only commercially available models. In Figure 5c, the most realistic scenarios are compared, based on the averaged data of Table 1. For hydrogen (P2H), the range shown is for small and large containers, where heat loss (and therefore liquid-loss) depends on the size-dependent surface-to-volume ratio.

As shown in Figure 2, when two curves (or lines) intersect, it can be seen that for storage shorter than the time corresponding to the intersection point, where one is the more energetically advantageous solution for storages involving shorter times, while the other is better for longer storage times. As shown in Figure 5c, certain types (gravitational storage, offline vanadium redox flow battery) are very advantageous for long term energy storage; their disadvantages related to other criteria will be discussed in the next section. Also, lead-acid batteries and second-life Li-ion batteries (at least the better ones) seem to be a good solution; half of the energy stored during the summer can be recovered after 2.5–5 months. A storage system supplying a Hungarian municipality of 10,000 inhabitants (based on energy demand of 4260 kWh/person/year) for three winter months is 10.65 GWh; it would be difficult to build a storage system of this size with these types of storage (numbers would be similar for most of the countries). For Li-ion batteries, the main limitation is the amount of lithium needed. This problem is further escalated by the fact that, unlike many other types, Li-ion batteries are also well-suited to transport applications, where they are in high demand leaving less batteries for utility-scale storage. Additionally, the relatively short lifetime of Li-ion batteries (<10 years, even with the second-life period) makes this type hardly applicable for multi-annual storage. For lead batteries, the potential environmental hazards would perhaps be the primary reason not to build such a storage facility.

Red dots indicate the time limits when P2M storage will be than these batteries. This occurs after around 118 days for 33% recovery (P2M-33%) compared to Li-ion batteries and after about 205 days compared to acid lead batteries; these values change to 76 and 138 days for 50% recovery (P2M-50%). In other words, for seasonal energy storage, where storage would mostly occur in July-August and use in December-February, i.e., 100–200 days later (electricity would have to be stored in an unloaded state until then), P2M method is already competitive with most other storage methods even at the 33% total storage efficiency that is currently easily achievable; the two exceptions to the types discussed are the gravitational storage and the offline vanadium redox flow battery. A comparison with these methods is the subject of the next sub-section.

The other types of storage considered (NaS battery, circulating VRFB battery and hydrogen storage with both liquid and gas storage) are not suitable for seasonal, annual or multi-annual storages.

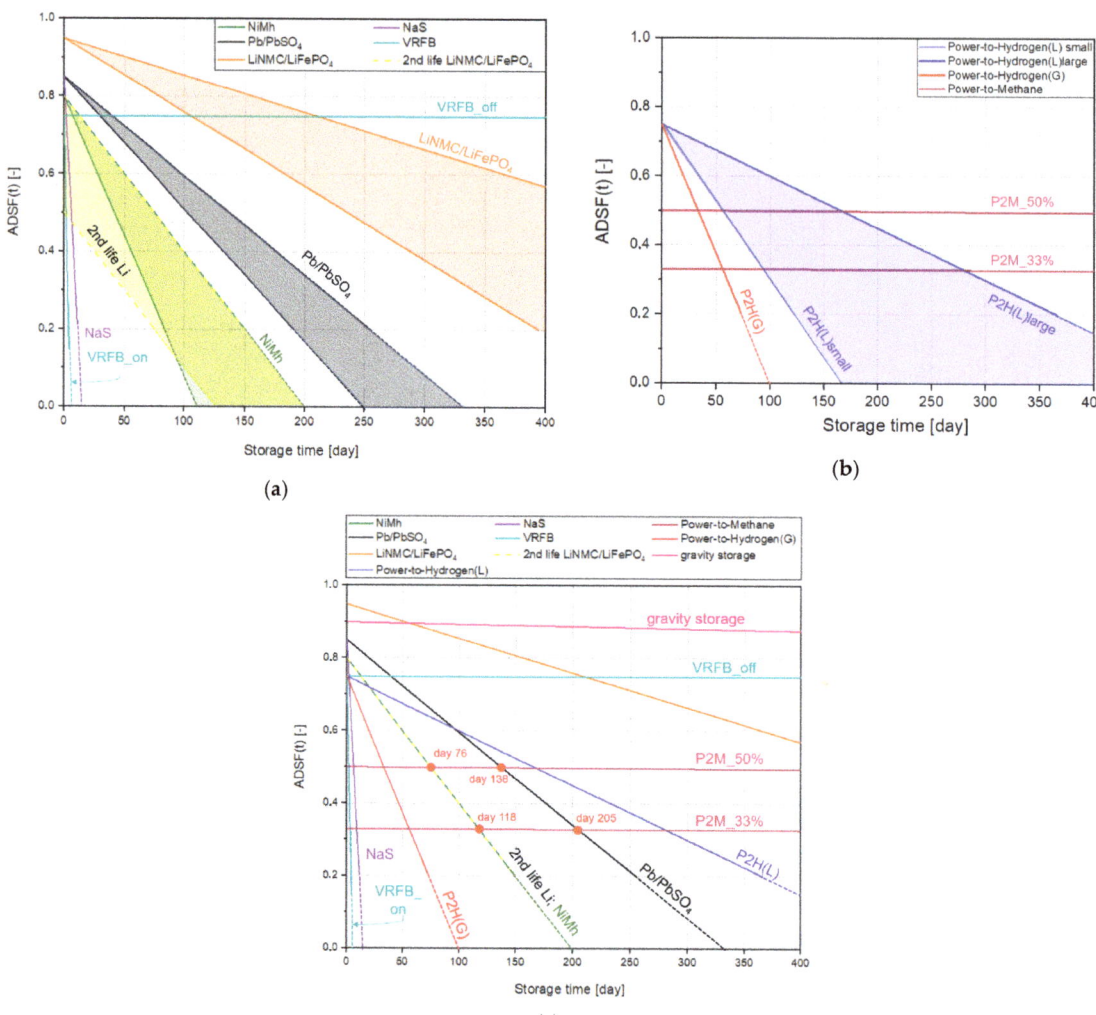

Figure 5. Comparison of simplified (linear) $ADSF(t)$ functions of the relevant storage methods. (**a**): Comparison of various battery types (marking the best and worst scenarios); (**b**): Similar diagram for the available Power-to-Gas methods; (**c**): comparison of the methods using data of the best commercially available models. The intersection points indicate which storage method is energetically better for storage times shorter than the corresponding duration or longer than the corresponding no-load storage time. The four red dots indicate the times for which the Power-to-Methane method may be energetically good for longer duration storage.

3.4. Comparison of High-Capacity Storage Solutions Applicable for Long-Time Storage

Based on the previous calculations, for seasonal to multi-annual storage, the Power-to-Methane method has two competitors, the gravity storage and the offline VRFB, where the liquid electrolyte tanks are separated from the central cell. Therefore, self-discharge is reduced practically to zero. The offline VRFB will not be competitive; vanadium is even more difficult to obtain than lithium, so for the 10 GWh demand of the town mentioned above, the production of 400,000 m^3 of solution with a concentration of at least 1 mol/L vanadium compound at a volume energy density of 25 Wh/L does not seem realistic.

A more serious challenge is the solid gravity storage, like mass- (or weight-) towers. Gravity storage is similar in principle to pumped storage in that it stores energy in the form of potential energy and can be recycled with high efficiency. Although there are some types that can only be installed in certain locations (e.g., hillsides [24] or mines [25]), energy tower-type versions [26] could be installed almost anywhere. In the energy tower, concrete blocks are stacked using a special crane; in this way, the electricity turns to potential energy. When discharged, the crane lowers these blocks to ground level, while a suitable transmission system generates electricity with a built-in generator. Self-discharge is difficult to understand in such storage, although in the long term, the falling or erosion of the stored blocks may cause such a loss. Since no numerical data were available, we looked for a long-standing tall structure built of heavy blocks and used it to investigate the reduction in stored energy over a sufficiently long period of time; from this, we were able to estimate the daily η_s value.

The studied "solid gravity storage" system is the Cheops pyramid; its original height is estimated at 146.7 m, now 138.8 m. Its current mass is about six million tons, its volume 2.3 million cubic meters, and its age is about 4500 years. We approximated both the original and the current version as a regular pyramid; the size of the bases was assumed to be constant, and the loss was calculated from the loss of mass and height. Therefore, the energy stored was about 634 MWh, and the current energy content was 567 MWh, a "self-discharge" of 10.5% over the whole lifetime, which is 6.4×10^{-6} per day, or 0.00064%/day, practically comparable to P2M methods, but the big advantage is the long lifetime in the "no load" condition, which in this case exceeds 1000 years.

A serious physical disadvantage of this type (the financial side is not considered in this article) is the large size due to the low energy density. With a medium-quality gas turbine, it would require about 75 tons of methane (natural gas) to produce the energy stored in such a gravity storage system. This is 0.00125% of the pyramid by mass, which is about 170 m^3 in liquid storage (LNG), about 420 m^3 in high-pressure storage (CNG, 200-250 bar) and about 100,000 m^3 at normal pressure. In other words, a Cheops pyramid-sized atmospheric pressure reservoir would have a storage capacity as a P2M reservoir of about 23 times that of a gravity reservoir, shifting this ratio even more at higher pressures; moreover, with P2M method, a pyramid would not need to be dismantled and built seasonally.

Power-to-Methane technology appears to be the best technical solution for seasonal, annual and multi-annual storage of large amounts of energy. It is important to note that this is an economically and socially acceptable method, which also fits well with the existing storage and electricity generation infrastructure [27–29].

Our aim was to show that there is a segment in the long-time (seasonal to multi-annual) energy storage, where Power-to-Methane technology can outperform other methods. This conclusion is valid only in the given storage-time range; for shorter or longer storage periods, other methods are better choices than P2M.

One of the main objection against P2M method is its relative un-maturity, compared to other storage technologies, like Li-ion batteries or even the other Power-to-Gas technology, the Power-to-Hydrogen method. In some sense, it is undoubtedly true that these methods are more established, but still, P2M technology is also notably an established method. In relation to hydrogen-based storage, water electrolysis can be considered a more established technology, but methanation—even the biochemical one—can also be considered a mature technology. This can be proved by the growing number of industry-scale biomethanation facilities, mentioning only a few of them, like MicrobEnergy—BioPower2Gas in Allendorf, Germany; the Electrochaea—BioCat in Avedøre, Denmark or the Underground Sun Storage in Pilsback, Austria.

4. Conclusions

Most of the currently used energy storage methods, which can store large amounts of energy, are used to compensate for the difference between fluctuating energy production and fluctuating demand. Battery systems are suitable for this purpose up to a few days period, even for larger quantities (e.g., a few MWh). However, for seasonal and even longer (annual to multiannual) storage, these types are not suitable.

In this article, a novel function has been introduced, shoving properties similar to the lifetime and efficiency. This Actual Discharge State Function ($ADSF(t)$) indicates the fraction of the energy which can be recovered from the storage system after a given unloaded period of time (t). This quantity is somewhat similar to the shelf-life quantity of batteries, but it does not indicate how long the storage device can be used, rather how long the stored energy can be used, with a certain recovery efficiency.

The following storage methods have been compared: lead-acid battery; nickel-metal hydride battery, lithium-ion (LiNMC/LiFePO$_4$) battery, vanadium redox flow battery (standby and offline modes), sodium-sulphur battery, Power-to-Hydrogen method (with hydrogen stored as pressurized gas or cryogenic liquid), Power-to-Methane method (with 33 and 50% recovery efficiency), and solid gravity storage systems (mass-towers). For seasonal energy storage, the P2M method can return the stored energy with higher efficiency than all other methods, except for VRFB with separated tanks (i.e., in offline mode) and the mass-tower storage. In relation to other technical criteria (such as size or availability of the necessary materials), P2M technology is superior to the other two methods and can therefore play an important role for seasonal (electricity will be generated in Summer, stored in the gas grid for a few months, then convert back to electricity in Winter) or longer (e.g., a few years) storage periods. On this basis, the P2M method can be positioned as a seasonal or multi-annual, high energy, relatively small (compact) energy storage system that can be "discharged" very easily and with acceptable efficiency.

Choosing the best energy storage solution for a given problem is a multi-dimensional optimization problem, where some of the functions to be considered are not even technical ones, but rather financial or even sociological. The function defined here can be used as one of the "technological" dimensions, but other dimensions have to be also considered; some of them with smaller, but others with bigger weight.

Supplementary Materials: The following are available online at https://www.mdpi.com/article/10.3390/en14113265/s1, Table S1: Shelf-lifes and range of constants of the simplified (linear) $ADSF(t)$ function (Equation (3)), used in Figure 5a,b.

Author Contributions: Conceptualization, A.R.I.; formal analysis, K.K. and A.R.I.; investigation, K.K. and A.R.I.; writing—original draft, K.K. and A.R.I. All authors have read and agreed to the published version of the manuscript.

Funding: This work was performed in the frame of the 2020-3.1.1-ZFR-KVG-2020-00006 project, implemented with the support provided from the National Research, Development and Innovation Fund of Hungary, financed under the 2020-3.1.2-ZFR-KVG funding scheme. Part of the research reported in this paper and carried out at BME has been supported by the NRDI Fund (TKP2020 NC, Grant No. BME-NC) based on the charter of bolster issued by the NRDI Office under the auspices of the Ministry for Innovation and Technology. K.K. has been supported by the EK2 Student Program of the Centre for Energy Research.

Conflicts of Interest: The authors declare no conflict of interest.

References

1. Directive (EU). 2019/944 of the European Parliament and of the Council of 5 June 2019 on Common Rules for the Internal Market for Electricity and Amending Directive 2012/27/EU (Text with EEA Relevance.) Chapter 1, Article 2, Point 59. Available online: https://eur-lex.europa.eu/legal-content/EN/TXT/?uri=CELEX:32019L0944 (accessed on 20 April 2021).
2. Huggins, R.A. *Energy Storage—Fundamentals, Materials and Applications*, 2nd ed.; Springer: Cham, Switzerland, 2016; ISBN 978-3-319-21238-8. [CrossRef]

3. Sterner, M.; Stadler, I. (Eds.) *Handbook of Energy Storage—Demand, Technologies, Integration*; Springer: Berlin, Germany, 2019; ISBN 978-3-662-55503-3. [CrossRef]
4. Györke, G.; Deiters, U.K.; Groniewsky, A.; Lassu, I.; Imre, A.R. Novel Classification of Pure Working Fluids for Organic Rankine Cycle. *Energy* 2018, *145*, 288–300. [CrossRef]
5. Hiesl, A.; Ajanovic, A.; Haas, R. On current and future economics of electricity storage. *Greenh. Gases Sci. Technol.* 2020, *10*, 1176–1192. [CrossRef]
6. Altun, A.F.; Kilic, M. Thermodynamic performance evaluation of a geothermal ORC power plant. *Renew. Energy* 2020, *148*, 261–274. [CrossRef]
7. Tamakoshi, T. Development of Sodium Sulfur Battery and Application. *Grand Renew. Energy Proc.* 2019, *2018*, 286. [CrossRef]
8. Farahani, S. *Battery Life Analysis in: ZigBee Wireless Networks and Transceivers*; Elsevier-Newnes: Oxford, UK, 2008; Chapter 6; pp. 207–224, ISBN 978-0-7506-8393-7. [CrossRef]
9. Benato, A.; Macor, A. Biogas Engine Waste Heat Recovery Using Organic Rankine Cycle. *Energies* 2017, *10*, 327. [CrossRef]
10. Macchi, E.; Astolfi, M. *Organic Rankine Cycle (ORC) Power Systems: Technologies and Applications*; Elsevier-Woodhead Publishing: Duxford, UK, 2016.
11. Ter-Gazarian, A.G. *Energy Storage for Power Systems (IET Power and Energy Series)*, 2nd ed.; The Institution of Engineering and Technology, IET: London, UK, 2011.
12. McIntyre, O. *Gabčíkovo—Nagymaros Project: A Test Case for International Water Law?* Anton, E., Anders, J., Joakim, Ö., Eds.; Transboundary Water Management: Principles and Practice, Stockholm International Water Institute; Routledge: London, UK, 2010; p. 228.
13. Martinez-Laserna, E.; Gandiaga, I.; Saraseketa-Zabala, E.; Badeda, J.; Stroe, D.-I.; Swierczynski, M.; Goikoetxea, A. Battery second life: Hype, hope or reality? A critical review of the state of the art. *Renew. Sustain. Energy Rev.* 2018, *93*, 701–718. [CrossRef]
14. Oshima, T.; Kajita, M.; Okuno, A. Development of Sodium-Sulfur Batteries. *Int. J. Appl. Ceram. Technol.* 2004, *1*, 269–276. [CrossRef]
15. Olabi, A.G.; Onumaegbu, C.; Wilberforce, T.; Ramadan, M.; Abdelkareem, M.A.; Al-Alami, A.H. Critical review of energy storage systems. *Energy* 2021, *214*, 118987. [CrossRef]
16. *EPRI-DOE Handbook of Energy Storage for Transmission &Distribution Applications*; EPRI: Palo Alto, CA, USA; The U.S. Department of Energy: Washington, DC, USA, 2003; p. 1001834. Available online: https://www.sandia.gov/ess-ssl/publications/ESHB%201001834%20reduced%20size.pdf (accessed on 1 February 2021).
17. Sperling, D.; Cannon, J.S. *The Hydrogen Energy Transition: Cutting Carbon from Transportation*; Elsevier: San Diego, CA, USA, 2004.
18. Kovac, A.; Paranos, M.; Marcius, D. Hydrogen in energy transition: A review. *Int. J. Hydrogen Energy* 2021, *46*, 10016–10035. [CrossRef]
19. Roensch, S.; Schneider, J.; Matthischke, S.; Schluter, M.; Goetz, M.; Lefebvre, J.; Prabhakaran, P.; Bajohr, S. Review on methanation—From fundamentals to current projects. *Fuel* 2016, *166*, 276–296. [CrossRef]
20. Hidalgo, D.; Martín-Marroquín, J.M. Power-to-methane, coupling CO_2 capture with fuel production: An overview. *Renew. Sustain. Energy Rev.* 2020, *132*, 110057. [CrossRef]
21. Vera, D.; Baccioli, A.; Jurado, F.; Desideri, U. Modeling and optimization of an ocean thermal energy conversion system for remote islands electrification. *Renew. Energy* 2020, *162*, 1399–1414. [CrossRef]
22. KSH—Methane (CH4) Emission of Varoous (Industry, Transportation, Households, etc.) Sources (Nemzetgazdasági Ágak és Háztartások Metán (CH4) Kibocsátása) (1985–)(4/4). Available online: http://www.ksh.hu/docs/hun/xstadat/xstadat_eves/i_ua028d.html (accessed on 1 March 2021).
23. Kirchgessner, D.A.; Lott, R.A.; Cowgill, R.M.; Harrison, M.R.; Shires, T.M. *Estimate of Methane Emissions from the U.S. Natural gas industry—2019*; U.S. Environmental Protection Agency. Available online: https://www.epa.gov/natural-gas-star-program/estimates-methane-emissions-segment-united-states (accessed on 1 February 2021).
24. Cava, F.; Kelly, J.; Peitzke, W.; Brown, M.; Sullivan, S. Advanced Rail Energy Storage: Green Energy Storage for Green Energy. In *Storing Energy—With Special Reference to Renewable Energy Sources*; Trevor, M.L., Ed.; Elsevier: Amsterdam, The Netherlands, 2004; Chapter 4; pp. 69–86.
25. Gravitricity—Gravity Energy Storage. Available online: https://gravitricity.com/ (accessed on 30 January 2021).
26. Moore, S.K. The Ups and Downs of Gravity Energy Storage: Startups are pioneering a radical new alternative to batteries for grid storage. *IEEE Spectr.* 2021, *58*, 38–39. [CrossRef]
27. Csedő, Z.; Sinóros-Szabó, B.; Zavarkó, M. Seasonal Energy Storage Potential Assessment of WWTPs with Power-to-Methane Technology. *Energies* 2020, *13*, 4973. [CrossRef]
28. Pintér, G. The Potential Role of Power-to-Gas Technology Connected to Photovoltaic Power Plants in the Visegrad Countries—A Case Study. *Energies* 2020, *13*, 6408. [CrossRef]
29. Pörzse, G.; Csedő, Z.; Zavarkó, M. Disruption potential assessment of the power-to-methane technology. *Energies* 2021, *14*, 2297. [CrossRef]

Review

Past, Present and Near Future: An Overview of Closed, Running and Planned Biomethanation Facilities in Europe

Máté Zavarkó [1,2], Attila R. Imre [3,4,*], Gábor Pörzse [5] and Zoltán Csedő [1,2]

[1] Department of Management and Organization, Corvinus University of Budapest, 1093 Budapest, Hungary; mate.zavarko@uni-corvinus.hu (M.Z.); zoltan.csedo@uni-corvinus.hu (Z.C.)
[2] Power-to-Gas Hungary Kft, 5000 Szolnok, Hungary
[3] Department of Energy, Faculty of Mechanical Engineering, Budapest University of Technology and Economics, Műegyetem rkp. 3, 1111 Budapest, Hungary
[4] Centre for Energy Research, Department of Thermohydraulics, Konkoly Thege Str. 29-33, 1121 Budapest, Hungary
[5] Corvinus Innovation Research Center, Corvinus University of Budapest, 1093 Budapest, Hungary; gabor.porzse@uni-corvinus.hu
* Correspondence: imreattila@energia.bme.hu

Abstract: The power-to-methane technology is promising for long-term, high-capacity energy storage. Currently, there are two different industrial-scale methanation methods: the chemical one (based on the Sabatier reaction) and the biological one (using microorganisms for the conversion). The second method can be used not only to methanize the mixture of pure hydrogen and carbon dioxide but also to methanize the hydrogen and carbon dioxide content of low-quality gases, such as biogas or deponia gas, enriching them to natural gas quality; therefore, the applicability of biomethanation is very wide. In this paper, we present an overview of the existing and planned industrial-scale biomethanation facilities in Europe, as well as review the facilities closed in recent years after successful operation in the light of the scientific and socioeconomic context. To outline key directions for further developments, this paper interconnects biomethanation projects with the competitiveness of the energy sector in Europe for the first time in the literature. The results show that future projects should have an integrative view of electrolysis and biomethanation, as well as hydrogen storage and utilization with carbon capture and utilization (HSU&CCU) to increase sectoral competitiveness by enhanced decarbonization.

Keywords: biomethanation; power-to-methane; competitiveness; hydrogen utilization; decarbonization; Hungary

1. Introduction

In line with the long-term strategy of the European Union to become climate-neutral, the energy storage challenge [1] that is induced by volatile renewable electricity production (e.g., with rapidly growing photovoltaic capacities) must be handled [2–4]. Power-to-gas (P2G), and especially power-to-methane (P2M), technologies, however, are capable of providing flexibility [5] and efficient seasonal energy storage [6] with the reuse of CO_2 and the utilization of the existing capacities of the natural gas grid [7]. Moreover, these technologies are not only present on a lab-scale or prototype level, but there are examples for commercial-scale implementation, with chemical [8] and biological methanation [9] as well. Widespread utilization of this technology, however, has not happened yet, despite the potential of P2G technologies [10,11]. To accelerate the implementation of the P2M technology on a commercial scale, further R&D&I activities and policy regulations are also needed [7]. Regarding the prior literature in the P2G field, the "research" and the "development" part of the R&D&I are often supported by new technoeconomic research results [12–14]. Moreover, the "innovation" part is already discussed from in-depth management aspects [7,15], and there are also analyses from policy perspectives [6,16]. Nevertheless, there is a lack of a high-level approach which can integrate these aspects for socioeconomic progress. Consequently, this study focuses on P2M facilities with novel biological methanation technology [17] and their potential connection to sectoral competitiveness in Europe.

Compared to previous project reviews [17–20], which have already collected P2M projects including chemical and biological methanation (and other P2X projects, as well), this study has a different approach with the following adjustments:

1. Narrowing the technological scope for biological methanation to generate a specific analysis;
2. Following a novel abductive methodological approach in this area with (1) using quantitative and qualitative data, (2) starting the analysis through the lens of a technology developer company, and (3) iteration with former theories and results to identify trends and gaps which can define the scope of future facilities;
3. Considering specific contributions of future projects to sectoral competitiveness in Europe.

This competitiveness-oriented approach is unique in the P2G literature. Even though Brunner et al. [21] analyzed the relationships of competitiveness and P2G, it had a different scope: they aimed to compare the competitiveness of different P2G operational concepts. Moreover, research usually focuses on the competitiveness of P2G technologies (e.g., compared to other energy storage technologies) [22,23] but rarely on the competitiveness-increasing opportunities by P2G (or P2M in this study). The importance of this topic, however, derives from the practical need and the context as well—similar to the competitiveness studies in general. For example, Fagerberg [24] argues that the "competitiveness" term also does not originate from theoretical researchers but professionals working around decision makers. The relevance of this topic has similar roots: the European Green Deal mentions several times the importance of supporting the economic competitiveness of the EU [25]. The document also declares that "new technologies, sustainable solutions and disruptive innovation are critical to achieve the objectives of the European Green Deal" (p. 18, [25]). As recent research focusing on biomethanation technology concluded that P2M can be disruptive in the future [15], the research question of this study is the following:

How can future P2M, and especially biomethanation facility development projects, increase sectoral competitiveness in Europe?

Figure 1 summarizes (1) why the research is relevant, (2) what is in the focus of the research, and (3) how the research was conducted.

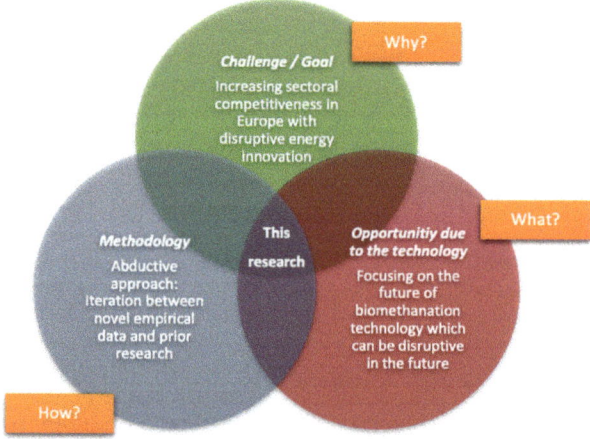

Figure 1. Relevance and scope of the research.

In sum, the study has a more future-oriented approach rather than a retrospective one, and this future orientation requires specificity regarding:
1. the technology (biomethanation);
2. the goal (supporting sectoral competitiveness by this technology);
3. the method (starting the investigation from the aspect of a concrete market player who may contribute to these goals).

Based on the abductive approach of this research with qualitative elements, hypotheses cannot be made, but underlying presumptions as propositional knowledge emerging from prior research [26] can be explicated that will be extended, modified, or developed further by empirical data gathering, analysis, and theory generation. Regarding the fundamental characteristics of the focal technology listed in the first paragraph, the underlying presumption for the research question is that future biomethanation facility development projects would increase sectoral competitiveness in Europe by providing flexibility, seasonal energy storage, and reuse CO_2 for synthetic natural gas production, thus integrating renewable energy sources and contributing to decarbonization efforts.

The study is structured as follows. First, the technical background of biological methanation, the research framework, and the applied data gathering and analysis methods are presented. In the Results section, biomethanation facilities are presented and key topics for future projects are revealed. After that, these topics are discussed in-depth according to former literature and research results. Finally, conclusions, limitations, and further research directions are outlined.

2. Materials and Methods

2.1. Technical Background

The study focuses on the methanation segment of P2G. In this case, the mixture of H_2 (from water electrolysis by renewable electricity) and CO_2 (e.g., from biogas, landfill gas or flue gas) can be converted to methane [27]: $CO_2 + 4H_2 \rightarrow CH_4 + 2H_2O$.

There are four different solutions for methanation, two of them are already in use at the commercial scale: chemical and biological methanation. These solutions have different operational characteristics that are thoroughly analyzed in the literature (see, e.g., [19] or [28]). One of the main differences is that while chemical (catalytic) methanation needs high pressure and temperature to reach high CO_2 conversion, which can be 50–60% and 80–90% or higher in proper conditions, biological methanation needs lower pressure and temperature (ca. 60–70 °C) than catalytic methanation; moreover, the CO_2 conversion is often higher than 95% [17,29–32]. Furthermore, in the case of biological methanation, microorganisms catalyze the reaction in a multiphase system because the gaseous H_2 and CO_2 are dissolved in the liquid phase, in which the Archaea absorb them and produce CH_4 [33]. In contrast, chemical methanation needs other types of catalysts, e.g., ruthenium [34] or nickel-based catalysts, that are characterized by high performance and low cost [35]. The catalysts determine different opportunities and limitations as well. For example, fluctuations and impurities are less harmful in case of the biological process with the robust methanogens, thus it can provide simpler applicability in contexts where contaminants (e.g., hydrogen sulfide) must be considered; nevertheless, its main limitation is the gas-to-liquid mass transfer at a relatively low temperature [19].

The "biomethanation" and "biological methanation" terms are often used in case of biogas upgrading, as well, when additional hydrogen injection happens. In this case, hydrogenotrophic methanogens function as a catalyst in a mixed culture, and there is no need for a separate bioreactor and clear culture [36,37]. A novel method for the P2M process is the bioelectrochemical system for electromethanogenesis (EMG-BES). It uses electro-active microorganisms, and the reaction happens only at 25 to 35 °C [38].

In the case of the focal biomethanation technology, microorganisms can convert ca. 97–98% of the CO_2 into methane during the methanation phase in a separate culture, which is promising regarding the decarbonization efforts. The total efficiency of such a biomethanation plant (together with the electrolysis step) can be in the range of 55 to 60% [6].

2.2. Research Framework

While competitiveness is defined from several aspects in the literature [39], an innovation approach must be considered in this study. In this sense, sectoral competitiveness can mean such capabilities which are (partly) created by innovation and which are required for sustained economic growth in an (international) competitive environment [40]. As innovation can be interpreted as a process during which an opportunity becomes a useful solution in practice [41] and can be a positive-sum game because of the complementarities among contributors [42], a network approach can be also important to increase sectoral competitiveness.

Accordingly, as the literature highlights the importance of the know-how transfer among companies, universities, and state administration [40,43], this research not only focuses on biomethanation projects but the main areas on which future work is necessary to contribute to sectoral competitiveness. For this purpose, the projects are interpreted from the aspect of recent scientific research results and EU strategies and policies. Figure 2 illustrates the research framework.

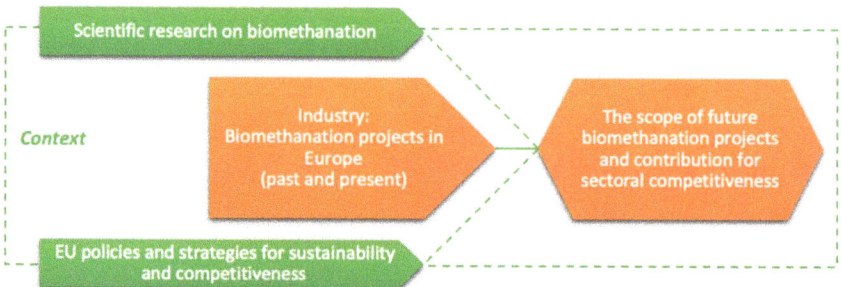

Figure 2. Research framework (Orange: Focus of the research; Green: Context of the research).

2.3. Research Methodology, Data Collection, and Analysis

An abductive approach was followed to answer the research question, which involved iteration between empirical data and theory. It means that empirical data can be analyzed to reveal regularities or phenomena, and then they can be compared to previous research results and theories to explain the revealed phenomena and develop new theories. This abductive approach and the iteration is emphatic in several research methods, such as grounded theory [44], extended case study [45], or more broadly, the abductive theory of method (ATOM) [46]. This research integrates elements from all of these methodological roots. As "ATOM itself as a grounded theory method that explicitly accommodates both quantitative and qualitative outlooks on research" (p. 106, [46]), this research involves both quantitative text analysis and their qualitative interpretation from the aspect of a disruptive technology developer company (an empirical case as a starting point).

The research is partly built on the digital R&D and open innovation platform of Power-to-Gas Hungary Kft., a Hungarian startup that developed an innovative biomethanation P2G prototype in cooperation with Electrochaea GmbH. The startup company consciously manages digital know-how flows within the organization and the inter-organizational network and continuously monitors the international P2G market and research results. On this platform, the company develops different kinds of knowledge elements with the involvement of employees, external professionals, stakeholders, and academic researchers. These knowledge elements include project descriptions, innovational and technological know-hows and analyses, e-learning materials, innovation problems, and ideas to solve them (idea generation). This platform, however, only represents the industrial "lens" for the study because research papers (indicating the scientific context) and EU policies and strategies (indicating the socioeconomic context) were also collected by the authors, while also considering the suggestions of the stakeholders of the company (interviewees, see below).

Table 1 presents the structure of data collection.

1. Focus: Relevant project descriptions were selected from the digital platform which meant direct benchmarking for the company for future project planning. These were collected based on the market monitoring activity by employees or suggestions by external professionals and other industry stakeholders. The selection criterion was that only European projects were within our scope.
2. Context:
    a. Scientific research can also affect the planning and implementation of future P2M projects with biological methanation [7]. The authors collected recent research papers and created a long list of potentially relevant publications from the aspect of the company and future biomethanation facility development ($n > 250$). The goal was to provide a broad horizon of opportunities (including power-to-methane (P2M), power-to-hydrogen (P2H), power-to-liquid (P2L), power-to-X (P2X), and carbon capture (CC)) and to avoid unintentionally narrowing the relevant themes, which could have limited the reliability of the research in an abductive sense. After that, the P2H-, P2L-, P2X, and CC-oriented contents were filtered out collaboratively with the interviewees. Facing the limited number of literature analyses focusing only on biological methanation, and as the term "biological methanation" is often used for novel biogas upgrading processes with H_2 injection and mixed culture, the potential contribution of biological P2M could be identified more reliably if less restrictions were applied and the whole P2M literature is considered ($n = 63$; see Appendix A).
    b. In line with the competitiveness approach of the study, the analysis of the project descriptions was compared to EU strategies for carbon-neutrality and their relations to competitiveness.

Table 1. Data collection for text analysis.

Data	Level of Analysis	Connection to the Research Framework	Relevance	Source/Suggested by	In Scope and Their Volume		Out of Scope (Examples)
Project descriptions	Micro	Focus	Review of industrial advancements	Employees, external professionals, and other industry stakeholders	Biomethanation projects in Europe (see the Results section)	21	Chemical methanation projects and/or out of Europe
Research papers, scientific publications	Meso	Context	Review of research directions and results	Employees and external academic researchers	P2M (see Appendix A)	63	P2H, P2L, P2X
Relevant policies and strategies	Macro		Outlining directions for technological innovations	EU websites	EU documents related to climate-neutrality policies and strategies and competitiveness: A Clean Planet for All: The European Green Deal [47–50]	4	Not EU or focusing on only economic competitiveness in general

The research involved quantitative text analysis with the JMP software, which is useful for text mining purposes [51]. After cleaning the data, recoding words and phrases (e.g., plurals, or "ptg" and "p2g" to "power to gas"), word clouds and trend analyses were generated, i.e., exploring the change of the most common terms according to different variables (time horizons, data sources (project descriptions scientific research or policies), and electrolyzer capacity in case of project descriptions). Moreover, trend analyses were combined with hierarchical clustering to reveal possible important underlying structures [52]. These quantitative analyses, additional qualitative interviews with company employees and stakeholders for interpreting raw data, and iterations with former literature indicated "key topics" for further elaboration. These key topics were iterated by more research results to generate an in-depth understanding and R&D&I directions for future biomethanation projects.

The relevance of these methodological choices is to look at the biomethanation projects not only through the lens of academia, but as a technology developer company as well. Power-to-Gas Hungary Kft. is known for its long-term mission to implement a 10 MW$_{el}$ P2M plant. It would be the largest P2M plant with biological methanation, and the second largest regarding chemical methanation as well [18]; consequently, the potential contribution to sectoral competitiveness is high in its case.

To improve the validity, reliability, and generalizability of the research, the following steps were undertaken:

1. Creating balance in authorship regarding research perspectives and background (energy research, applied research and development, technical aspects, economic and management aspects);
2. Building on the quantitative text analysis of more than 80 texts (more than 6000 total terms). The data sources had similar volumes regarding the number of terms (project descriptions: 2501; research abstracts: 2258; EU documents: 2341);
3. Triangulation—Involvement of professionals through interviews to support the interpretation of raw data and results;
4. Iteration between the literature and empirical data allowed us to develop conclusions that are valid in a specific context [44].

3. Results

3.1. Biomethanation Projects and Industrial-Scale Facilities in Europe

Regarding list of the European P2M projects with biological methanation, while most of the projects were listed in recent reviews of Thema et al. [19] and Bargiacchi [18], there are three projects that were not listed previously.

1. In contrast to the well-known biogas-based biomethanation projects, the BIOCO$_2$NVERT project aims to implement a biocatalytic P2M facility at one of the largest bioethanol plants of Europe. According to the description of the Innovation Land Lab, installation and commissioning are the next steps of the project [53]. The project started in 2018, and the cooperation partners are Klärgastechnik Deutschland GmbH, MicrobEnergy GmbH, PRG Precision Stirrer Gesellschaft GmbH, and Südzucker AG [54].
2. The HYCAUNAIS project takes place in Saint-Florentin, France, and involves synthetic methane production with CO_2 from landfill gas through the development of biological methanation. The project started in 2018 and is being realized by five private and three public partners [55,56].
3. The CarbonATE project in Austria and Switzerland focuses on the optimization of microbiological methanation by the development of enzymatic CO_2 capture process to prevent the microorganisms from harmful contaminants (e.g., N_2, O_2) of potential input gases (industrial exhaust gases) [57,58].

Table 2 shows these projects with the other projects which are monitored by the company based on accessible information about their capacity or status. Besides these 21 projects, Thema et al. [19] listed other biomethanation (mainly research) projects without sufficient (accessible) information:

1. "Biological biogas upgrading in a trickle-bed reactor" (Tulln/Donau, Austria, 2013);
2. "Biocatalytic methanation" (Cottbus, Germany, 2013);
3. "Forschungsanlage am Technikum des PFI" (Pirmasens, Germany, 2013);
4. "BioPower2Gas-Erweiterung" (Allendorf (Eder), Germany, 2016);

5. "Biologische Methanisierung in Rieselbettreaktoren" (Garching, Germany, 2016);
6. "Einsatz der biologischen Methanisierung [. . .]" (Hohenheim, Germany, 2016);
7. "Biocatalytic methanation of hydrogen and carbon dioxide in a fixed bed bioreactor" (Helsinki, Finland, 2016).

Regarding future industrial-scale developments, it can be argued that the capacity of larger biomethanation facilities must reach at least 1 MW_{el} to satisfy the demand for electrolysis and also for methanation, which can exceed even 1000 GW_{el} globally [59], and even over 500 GW_{el} for P2M in very positive scenarios [60]. Based on publicly accessible data, six projects reached or are planned to reach the 1 MW_{el} capacity: Energiepark Pirmasens-Winzeln, BioCat Project, Dietikon Microbenergy, INFINITY 1, Power-to-Gas Hungary plant, and HYCAUNAIS. The first five received a detailed description recently by Bargiacchi [18], while the HYCAUNAIS project was introduced above.

Table 2. European biomethanation projects with sufficient accessible information about capacity or status, based on [18,19] and own research.

Projects	Country	City	Start of the Project	Electrolyzer Capacity (MW_{el})	Status	Source of Status Information
PtG-Emden	Germany	Emden	2012	0.312	Closed	[18]
PtG am Eucolino	Germany	Schwandorf	2013	0.108	In operation	[61]
P2G-Foulum Project	Denmark	Foulum	2013	0.025	Closed	[62,63]
SYMBIO	Denmark	Lyngby	2014	-	Closed	[64]
W2P2G	Netherlands	Wijster	2014	0.400	In operation	[65]
BioPower2Gas	Germany	Allendorf	2015	0.300	Closed	[66,67]
GICON-Großtechnikum	Germany	Cottbus	2015	-	In operation	[68,69]
Energiepark Pirmasens-Winzeln	Germany	Pirmasens	2015	1.800	In operation	[70]
Mikrobielle Methanisierung	Germany	Schwandorf	2015	0.275	-	[71,72]
Biogasbooster	Germany	Straubing	2015	-	In operation	[73,74]
BioCat Project	Denmark	Kopenhagen/Avedore	2016	1.000	Closed	[75,76]
Power to Mobility (MicroPyros GmbH)	Germany	Weilheim-Schongau	2017	0.250	Under development	[77]
STORE&GO	Switzerland	Solothurn/Zuchwil	2018	0.350	Closed	[78]
ORBIT 1st site	Germany	Regensburg	2018	-	Closed	[79]
BIOCO$_2$NVERT	Germany	Dörentrup	2018	-	Under development	[53]
HYCAUNAIS	France	Saint-Florentin	2018	1.000	Under development	[55,56,80]
Dietikon Microbenergy	Switzerland	Dietikon	2019	2.500	Under development	[81]
ORBIT 2nd site	Germany	Ibbenbüren	2020	0.001	In operation	[82]
INFINITY 1	Germany	Pfaffenhofen a. d. Ilm	2020	1.000	Under development	[83]
CarbonATE	Austria and Switzerland	Winterthur	2020	-	In operation	[57]
Power-to-Gas Hungary plant	Hungary	-	-	10.000	In planning	[84]

3.2. Key Topics of Future Implementation

In the following section, key topics are identified based on the quantitative text analysis, interviews, and the literature, for which overarching R&D&I directions will be suggested in the Discussion section.

3.2.1. Key Topic 1: The Role of Biomethanation in the Hydrogen Economy

Based on the short summaries of the listed projects (ca. 1–3 pages, 2501 total terms, 14,573 total tokens), the most common word is "hydrogen". Similar influential words are "carbon dioxide" and "methane" (see Figure 3). This result refers to the importance of input factors in the biomethanation sector, and even though it is not particularly surprising, the relative dominance of hydrogen against the other key terms (e.g., methane, storage, biogas, natural gas) is conspicuous. Regarding the trend and advancements towards the hydrogen economy [85], the hydrogen orientation can be justified, but it can also be asked, for example, what could the role of biomethanation (biomethane or SNG production) be in the hydrogen economy? This might require further analysis later, based on other research results and EU policies.

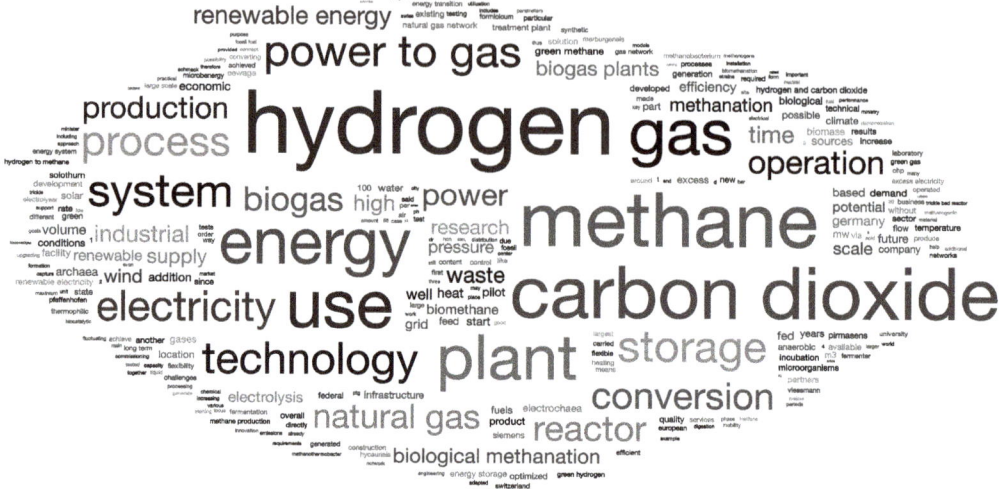

Figure 3. Word cloud from biomethanation project descriptions.

In addition to taking a static "snapshot" of the content of the biomethanation project descriptions (Key topic 1), quantitative characteristics of the projects provide opportunities for deeper insights. The next section (Key topic 2) considers the size of the facility (indicated by the capacity of the electrolyzer), while after that, Key topic 3 analyzes trends according to the start of the project (year).

3.2.2. Key Topic 2: Opening New Ways besides Biogas Plants to Store More Renewable Electricity/Hydrogen

The terms that appeared at least 15 times in the project descriptions were analyzed according to the size of the biomethanation facility. It can show how the focus of the R&D&I activities changes (or does not change) with the deployment of larger facilities. Considering the lessons of the interpreting interviews as well, Figure 4 suggests the following:

(a) at the small scale, the focus is on the "efficiency" of the "process", the "reactor" structure, the microorganisms ("archaea"), and the "biogas" input from "biogas plants", which contains "carbon dioxide" to "convert" it into "methane".

(b) at the large scale, the emphasis is on the "volume" of "wind" or other "renewable energy" and the "production" of "methane", which can be "fed" into the "natural gas" for "energy" "storage" purposes. (Words in quotation marks refer to the empirical data.)

The importance of the results shown by Figure 4 is that they interconnect the past and the future of biomethanation technology development from a purely technical aspect (without considering the time horizon, which is presented in the next section). Less abstractly, different issues are important at the small and large scales, and the gaps between these issues might generate new areas for research. Regarding the listed (a) and (b) points above, a step is missing between the efficient process in kW-scale with CO_2 from biogas and the purpose of storing high volumes of renewable electricity in the form of SNG in MW-scale. This missing step seems to be the sourcing of CO_2 in large volumes to develop multi-MW biomethanation plants. Accordingly, it is worth analyzing that if biogas plants cannot provide enough carbon dioxide for large-scale P2M plants [15], which could convert the vast volume of renewable electricity produced by wind or solar parks, what solutions can help to increase the capacity of biomethanation facilities to multi-MW$_{el}$ level, which are needed in the future [59].

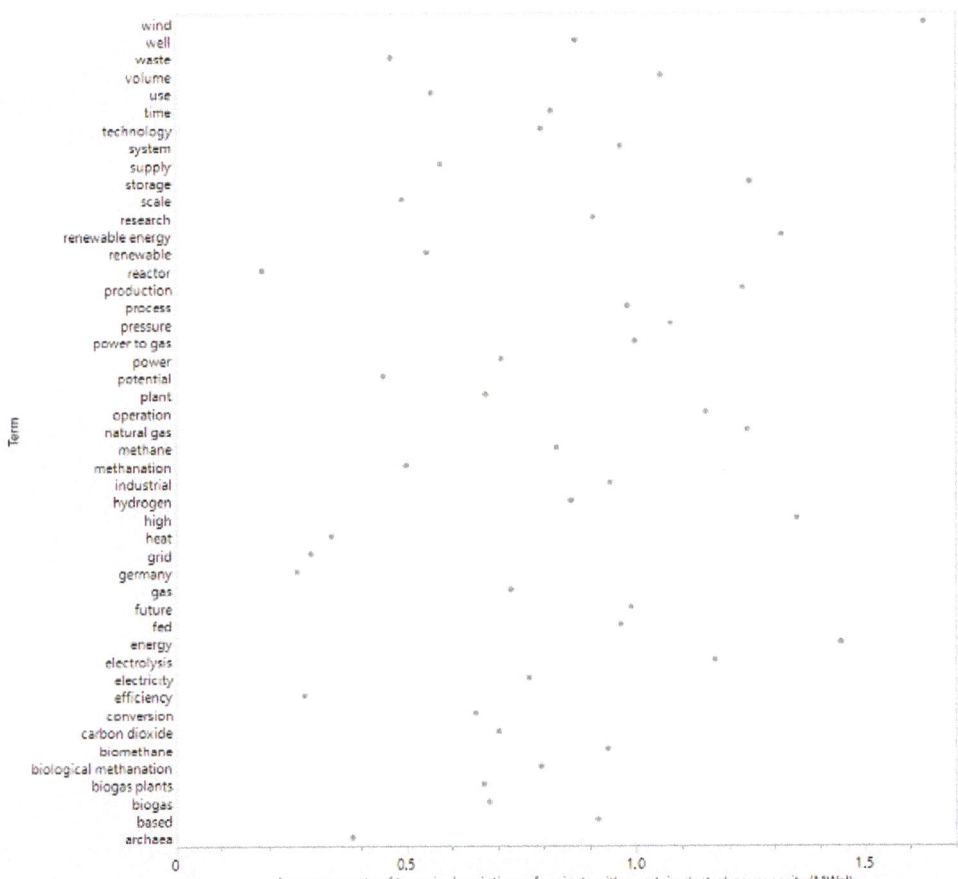

Figure 4. Appearance of the most common terms according to the size of the planned/implemented biomethanation facility. For example, the term "storage" appeared in project descriptions of biomethanation plants, which were 1.3 MW$_{el}$ on average.

3.2.3. Key Topic 3: From Technology Development towards Achieving "Future" Benefits

The most common terms of the project descriptions may change according to the start year of the projects, not only their capacity. Accordingly, Figure 5 shows constellation plots of a hierarchical cluster analysis which might reveal some underlying structures (e.g., main terms of past and present; based on the 75 most common terms). Based on the collaborative interpretation with the interviewees, Figure 5 shows the following:

1. From 2013 to 2016/2017, the emphasis was on "research" and "pilot" implementation; moreover, the fundamental characteristics of the process (e.g., using "excess" "solar" energy, "conversion" into "gas", connection to the "grid", and/or "biogas plants").
2. From 2016/2017 to 2020/2021, broader themes appeared, such as the "future" "potential" of the "technology" realized by a "company", utilizing "renewable energy" and

"electricity", and producing "green methane", "biomethane", or other "fuels" that fit the "infrastructure" to fight "climate" change.

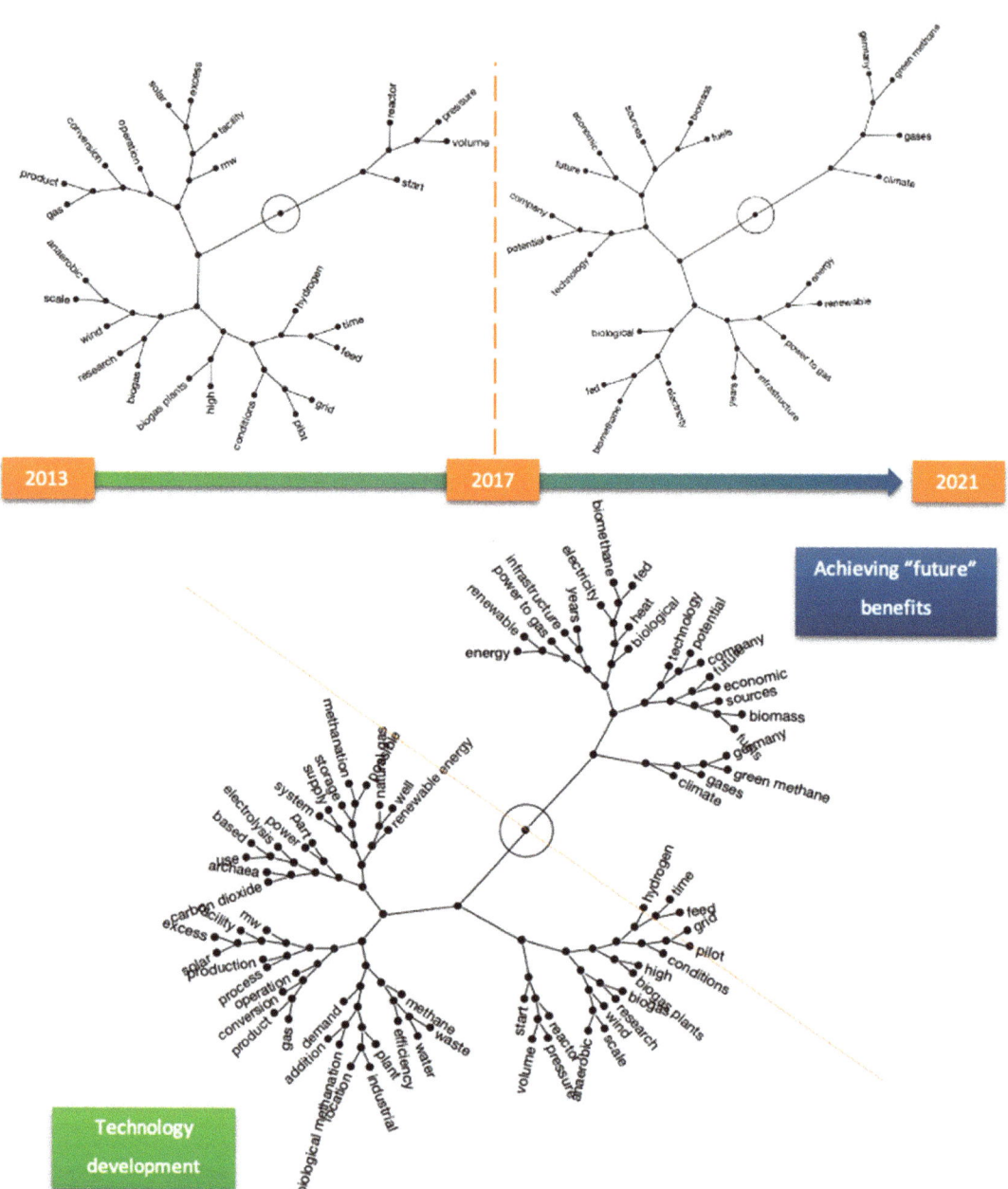

Figure 5. Constellation plots showing clusters of terms according to the project starts.

3.2.4. Key Topic 4: Future Project Planning in Line with Scientific Advancements and Policy Objectives

Based on the abstracts of the selected publications, a slightly different scheme can be seen on the word cloud than in the case of the biomethanation project descriptions. For example, while hydrogen and carbon dioxide are apparently important, carbon dioxide appears more often in case research papers, while hydrogen utilization appears more often in case of the project descriptions. Scientific research, however, deals more with the operative questions of the "system", the "process", or the "reactor", while biomethanation project descriptions write about "using" the "technology" for "energy storage" and the "production" of methane.

Figure 6 shows the comparison of the most common terms of research papers, project descriptions, and EU policies. In line with the mentioned trends, carbon dioxide (N = 116), hydrogen (101), methane (92), system (79), and power-to-gas (77) were the most dominant in a quantitative sense, in case of the abstracts of research papers. Project descriptions were also focused on these three terms, but with others: hydrogen (104), methane (90), carbon dioxide (80), use (80), energy (77), and gas (77).

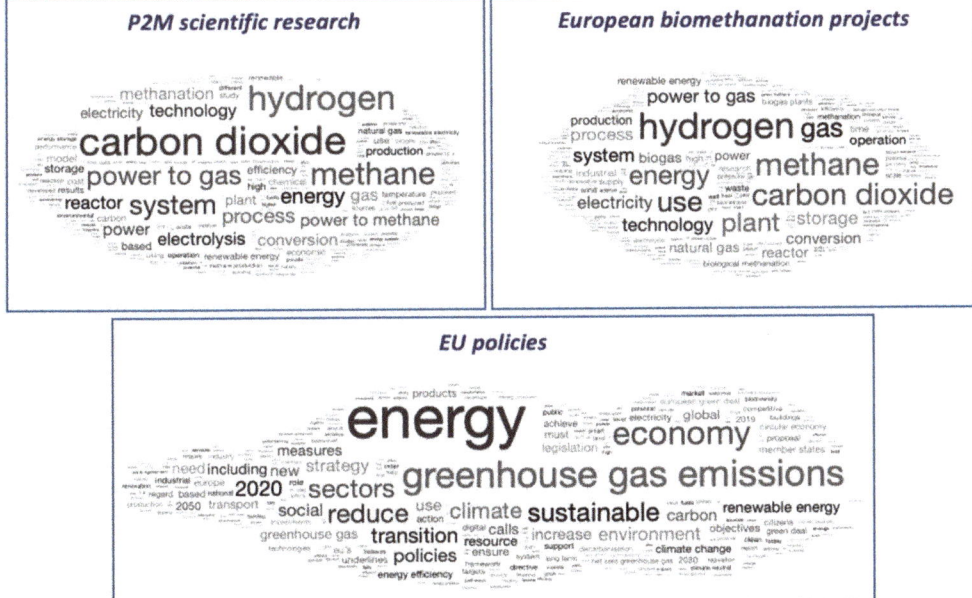

Figure 6. Comparing the word clouds of biomethanation project descriptions, P2M publications, and relevant EU policies.

In contrast to that, regarding the EU policies and strategies, the most common terms are energy (157), greenhouse gas emissions (88), economy (87), reduce (69), sectors (68), and sustainable (68). Accordingly, the main objective is to "reduce" the "greenhouse gas emissions" (GHG) through a "sustainable" "transition" with more "renewable energies". The "economy" and increasing "competitiveness" in a "global" "environment", however, is also important while fighting "climate change". For example, the document called "A Clean Planet for All" by the European Commission refers to competitiveness already in its subtitle: "A European strategic long-term vision for a prosperous, modern, competitive and climate neutral economy" [47]. The European Green Deal "aims to transform the EU into a fair and prosperous society, with a modern, resource-efficient and competitive economy" (p. 2, [48]).

These results suggest that scientific research, industrial project development, and policies have common points, e.g., the GHG-reduction induces the scientific research on carbon capture and utilization (CCU) solutions and their industrial application at biomethanation facilities. These high-level interconnections, however, should be analyzed in-depth to identify how sectoral competitiveness can be supported in practice by new biomethanation facility development projects.

4. Discussion

In the following, possible R&D&I directions will be presented for the identified key topics. These directions are based on the European context on the one hand, and the mentioned EU documents and the extensive studies of the European STORE&GO project [86] were analyzed. This project included three P2G demonstration plants, and its studies are heavily based on empirical data, as well.

On the other hand, the Hungarian context was also considered. It was important because the whole research followed an abductive approach in which the aspect of the Hungarian technology developer company is central. In addition, the capacity of the gas-grid of Hungary—compared to most other European countries—is quite high, which is an important factor to choose P2M-based long-term energy storage. Consequently, its national environment must be taken into account to contextualize findings. Since the National Energy Strategy 2030 of Hungary states that the implementation of the strategy contributes to the improvement of Hungary's competitiveness (p. 13, [87]) and identifies the direct value-creating potential of power-to-gas in energy storage (p. 39), the contribution for competitiveness by P2M seems clear by deductive reasoning. Based on the abductive methodology, however, it must be supported by more empirical evidence and prior research results.

In the following, overarching directions are presented along with the key topics for future biomethanation facilities, which can directly or indirectly increase sectoral competitiveness.

4.1. Key Topic 1: The Role of Biomethanation in the Hydrogen Economy—An Integrative View of Electrolysis and Biomethanation for Carbon-Neutral Energy Production, Flexibility Services, and Hydrogen Storage and Utilization

Hydrogen is explicitly considered as a "priority area" (p. 8, [48]) in the EU. The more hydrogen is produced in the hydrogen economy, the higher the need will be to store or utilize it efficiently, especially because there are safety limits to the injection hydrogen into the natural gas grid [88,89]. Biomethanation can function as a chemical method for hydrogen storing and/or utilization "tool" in the form of methane (SNG) in high amounts and for a long term, or it can be a middle step towards utilization in other forms (e.g., LNG). This is also in line with EU strategies, for example: "Sustainable renewable heating will continue to play a major role and gas, including liquefied natural gas, mixed with hydrogen, or e-methane produced from renewable electricity and biogas mixtures could all play a key role in existing buildings as well as in many industrial applications" (p. 8, [47]). Biomethanation can be also dynamically coupled with electrolyzers because microorganisms are capable to produce methane in seconds [6] (unlike chemical methanation)—this means additional flexibility beyond the electrolysis for the coupled electricity and gas sector.

In the European context, flexibility by electrolysis and biomethanation can have a beneficial effect on the operation of the network. The presence of 7.2 GW_{el} P2M could significantly reduce the peak (~45%) and duration (>95%) of the imbalance. Due to P2M installation, reverse energy flow could be reduced by 67% or even 100% [90]. Furthermore, the prevalence of green methane is also expected. By 2030, according to scenarios that can be considered optimistic in this regard, the 4% share could increase to 12% [91]. With P2M technology, the ratio of imported gas and total gas consumption could be reduced by up to 30–40% by 2050 [92]

In the Hungarian context, carbon-neutral methane production can also reduce natural gas imports, which is relevant because ca. 80% of the natural gas demand is covered by imports (p. 14). Encouraging the use of biogas, biomethane and non-natural-gas-based hydrogen (p. 12) also include the production of these energy sources. Hydrogen and methane can be used to produce biofuels or "green" fuels, which can be the part of the green transportation program (p. 13, [87]).

4.2. Key Topic 2: Opening New Ways besides Biogas Plants to Store More Renewable Electricity/Hydrogen—Enhanced Decarbonization by the Co-Specialization with Carbon Capture Technologies

Decarbonization is one of the main objectives in the EU, as "further decarbonising the energy system is critical to reach climate objectives in 2030 and 2050" (p. 6, [48]). In this area, an important challenge emerges in case of biomethanation because methanation requires efficiently useable CO_2 sources, which are reachable at biogas or bioethanol plants, but these plants sometimes do not have a close connection to the natural gas grid for energy storage, nor enough CO_2 for multi-MW_{el} P2M plants [6]. Moreover, carbon capture solutions for flue gas are considered expensive and immature in commercial-scale yet [15], despite the numerous research on different CC solutions (e.g., post-combustion [93] or oxyfuel-combustion [94]). A promising direction can be, however, the joint R&D on carbon capture technologies and biomethanation, similar to the research of Bailera et al. [95,96]. In addition to the co-specialization of resources, technologies can lead to competitive advantages in general in a competitive environment [97], and the context is also supporting this purpose. For example, by implementing P2M systems, pollutant emissions can be reduced, thus their environmental impact is positive and most of the positives can be detected in the field of climate change [98]. Synthetic methane offers outstanding greenhouse gas savings when biogenic carbon dioxide sources have been used in the methanation process or when hydrogen is generated by electrolysis by renewable energy [99], but it can be further be enhanced with efficient carbon capture solutions for flue gas.

Strictly speaking, P2M is not carbon-neutral because, during the use of biomethane, the previously "captured" CO_2 will be emitted again. However, one should realize, for example, that by using the flue gas of an energy-producing gas turbine for methanization and by reusing the new biomethane again in the same gas turbine, one carbon atom will be used twice (or thrice or even more) before emitting it as CO_2.

In the Hungarian context, by 2030, 90% of domestic electricity production is planned to be CO_2 emission-free (p. 42). Installation alongside GHG-intensive industrial activities and the use of industrial carbon dioxide are also promising in the methanation step, to increase the competitiveness of GHG-intensive industrial activities (p. 50). The production and purification of biogas can also contribute to the achievement of decarbonization goals (p. 20, [87]), and it is particularly true if carbon dioxide can be converted into methane.

4.3. Key Topic 3: From Technology Development towards Achieving "Future" Benefits—Finding Ways for Realizing and Communicating Business, Societal and Residential Value Creation

Research results suggest that even though there is still a need for research on biological methanation to increase its TRL [100], it is already worthwhile to analyze how future biomethanation can achieve future socioeconomic benefits. Realizing benefits is possible by scaling up the technology, but it requires capital investments, the returns of which might not meet the expectations of market players in the present [6,7]. Consequently, supporting regulations and profitable business models must be developed to "attract support from "patient" capital (i.e., long-term venture capital)" (p. 24, [47]), but significant public funding could be also necessary to scale up the technology. Public funding, however, can be justified if broad social and residential benefits are also explicit. Research results of the STORE&GO project show, for example, that a potential supply shock in the energy market will have less of an impact on social welfare if P2M technology is used [98]. Moreover, solar parks and P2M infrastructure increase the acceptance of the energy system by local energy communities [101].

In Hungary, rapidly growing photovoltaic capacities to 6000 MW by 2030 is a priority (p. 14), which can be technically supported by the integration function of P2M technology.

Converting surplus electricity to methane and its storage, however, can be important from broader socioeconomic aspects as well, because it can contribute to the affordable and steady energy supply (p. 10, [87]).

4.4. Key Topic 4: Future Project Planning in Line with Scientific Advancements and Policy Objectives—Building and Supporting Innovation Ecosystems for Efficient Know-How Transfer

Most of the biomethanation projects are realized in consortia of heterogeneous stakeholders, such as universities, startups, and large energy companies, while funding is often provided by the state administration [17,18]. As such, an inter-organizational network can have heterogeneous stakeholders, and an R&D&I ecosystem might be needed to bring and hold them together. An innovation ecosystem is a dynamic and adaptive system, the participants of which have different roles, motivations, and capabilities, but they all contribute to the success of an innovation process [102]. It means that an innovation ecosystem can involve not only companies from a certain sector but also from supporting sectors and regulators, since government policies also affect competitiveness [103]. Accordingly, the literature deals with the state support of R&D&I ecosystems as well. For example, government interventions that focus resources for grand challenges such as climate change facilitate the formation of networks beyond sectors and encourage scientific and technological improvements and the introduction of new or existing technologies [104]. These ecosystems can be the key for the economic growth after the COVID-19 pandemic according to the World Economic Forum, who also stated that supporting ecosystems can contain incentives for venture capital investments, R&D process and spreading new technologies [105]. Even though "green" R&D and creating balance among economic, ecological, and societal aspects must be supported by public financial sources, recent empirical studies showed that advancement in one dimension (e.g., ecological) does not necessarily happen at the cost of another dimension (e.g., economic) [106]. Biomethanation-focused innovation ecosystems may obtain support from the Horizon Europe programme, in which "partnerships with industry and Member States will support research and innovation on transport, including batteries, clean hydrogen, low-carbon steel making, circular bio-based sectors and the built environment" (p. 18, [49]).

Consequently, know-how transfer in biomethanation-focused innovation ecosystems with startups, universities, and state administrations can facilitate the previous three R&D&I directions. Figure 7 summarizes the findings and shows them according to the research framework.

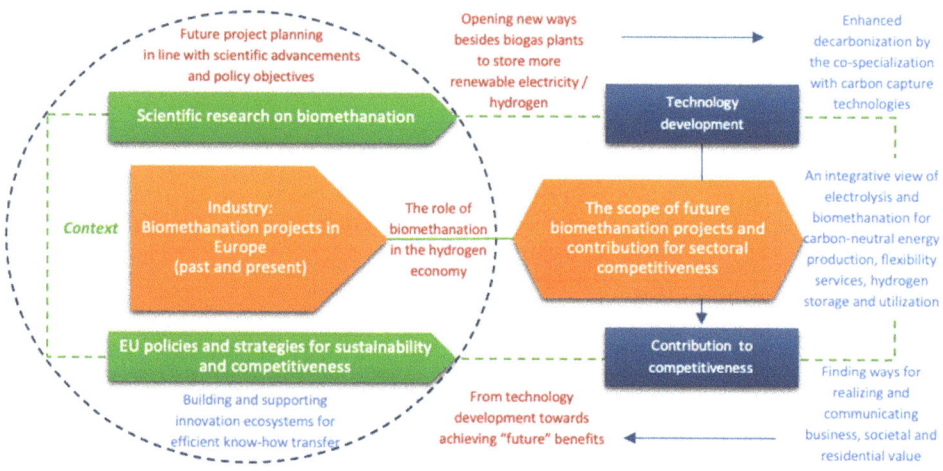

Figure 7. Findings aligned by the research framework (Orange: Focus of the research; Green: Context of the research; Red: Key topics; Blue: Suggested R&D&I directions).

4.5. Comparing the Results to Other Relevant Topics

As stated before, the list of the analyzed project descriptions and research papers was formulated from the viewpoint of a startup company (which might contribute to sectoral competitiveness with the deployment of large biomethanation plants). In addition, raw text data and interview data were iterated with former theories and research results. This approach, on the one hand, can be considered as a pragmatist one, which is in line with the roots of competitiveness studies [24], and one aimed to generate conclusions representing "cognitive usefulness in the world" (p. 18, [46]) and actionable knowledge [107]. This actionable knowledge is represented by the suggested R&D&I directions. Moreover, the conclusions would be supported by the coherence theories of truth, as well, according to which the propositions must be coherent with other scientific propositions [46]. This is because the empirical data was processed with comparisons with previous research results.

On the other hand, findings generated by this approach cannot cover every aspect and do not mean a positivist, general, theory [44]. It means that because of this abductive methodological approach, there can be areas that did not emerge during the research, i.e., the empirical data did not orient the parts of the ATOM, such as phenomena detection, analyses, and theory development [46]. Consequently, even though the synthesis of new empirical data and former research results into a coherent (but not full) set of theoretical propositions, it is worth comparing the results to some other topics which might be considered "key topics" in different contexts.

For example, (1) circular economy models can be relevant. Based on the analysis of Kircherrr et al. [108] on circular economy definitions, the main elements are central in circular economy models: the combination of reduction, reuse and recycling, and systemic shifts. These elements can be supported by biomethane production with P2M technology, as follows:

1. reuse of carbon dioxide happens in the methanation step;
2. the share of fossil energy sources can be reduced by the higher integration of renewable electricity and its storage in the form of methane;
3. coupling of the electricity and gas sectors means a system-level novelty, so the parallel function of energy storage and gas production of P2M.

Taking a closer look at biomethanation, research on the relationship with circular economy outlined some opportunities and challenges already. Baena-Moreno et al. [109] discussed that the combination of biological processes and renewable energy production can be the main pillar of the paradigm shift towards the circular economy, but incentives and/or cost-reduction-oriented technology developments are still needed. In a similar area, Eggemann et al. [110] argued that power-to-fuel processes producing methanol can contribute to the circular economy, but technology adoption might be influenced by the economic performance of these systems compared to other technological options. D'Adamo et al. [111] showed how biomethane can integrate effective management of renewable energies and municipal waste, thus contributing to the circular economy development. Their research leads to another important area of potential competitiveness developments, as well. The authors pointed out that biomethane can be used as fuel for vehicles, so (2) the green revolution in the transportation sector might be also supported by converting biowaste into clean fuels [111]. For example, biomethane can be compressed (CNG) or liquefied (LNG) for these purposes. Finally, competitiveness can be researched from (3) policy perspectives. Wall et al. [112] pointed out that recent EU legislation incites third-generation biofuels, and it creates a foundation for the integration of different solutions, e.g., anaerobic digestion, gasification, P2G, or algae as feedstock. Nevertheless, there is still further need for policy interventions to support green transitions with biomethane, for example, fueling stations in the case of the transportation sector [111], feed-in tariffs supporting seasonal energy storage [6], or carbon taxes [7] or certifications for green premiums [113] influencing the diffusion of the biomethanation technology and the production of biomethane.

These areas (circular economy, green transportation, impact of policies) did not emerge empirically during this research, but their related research results are mostly in line with the key topics and/or suggested R&D&I directions. For example, circular economy development is often focused on decarbonization which can be supported by the co-specialization of biomethanation and carbon capture technologies. Regarding green transportation, finding the role of biomethane in an area dominated by electric and hydrogen vehicles is also an important task, while incitive policies are important in the case of the formation of efficient national or international innovation ecosystems as well. The main novelty of the findings compared to previous literature, however, is related to the urged adaptation to the hydrogen economy and carbon capture of the future biomethanation project planning, as discussed below.

5. Conclusions

This research aimed to answer how could future biomethanation facility development projects increase the sectoral competitiveness in Europe. The propositional knowledge of the research based on prior literature was that future biomethanation facility development projects would increase sectoral competitiveness in Europe by providing flexibility, seasonal energy storage, and the reuse of CO_2 for synthetic natural gas production, thus integrating renewable energy sources and contributing to decarbonization efforts. Based on the empirical data collection, analysis, and iterative theory generation, however, this proposition must be modified. It viewed the relationship of biomethanation projects and sectoral competitiveness too narrowly, and ignored their contribution to the hydrogen economy and the synergies with another technology development area: carbon capture. The specific conclusion for this proposition is that in addition to the important energy storage potential, biomethanation facilities would increase sectoral competitiveness mainly due to their connective role between the two most important terms (or areas) of European strategies about economic and environmental progress: hydrogen economy and decarbonization.

Findings suggest that by building on the know-how of past and present projects, future biomethanation projects could take significant steps towards multi-MW_{el} capacities. Moreover, they should take these steps to satisfy the growing demand for their outputs and positive externalities. These improvements could support sectoral competitiveness in Europe if these projects have an integrative view of electrolysis and biomethanation for carbon-neutral energy production, flexibility of services, and hydrogen storage and utilization. In other words, biomethanation should be interpreted in the future as a hydrogen storage and utilization (HSU) solution, on the one hand. On the other hand, as biogas plants sometimes cannot provide enough CO_2 for multi-MW_{el} plants, enhanced decarbonization can be only achieved by co-specialization with carbon capture technologies. Consequently, future industry-scale biomethanation facilities should integrate hydrogen storage and utilization and carbon capture and utilization functions (HSU&CCU) to increase sectoral competitiveness in Europe. This direction, however, requires strategic alignment and know-how transfer among universities, startups, large energy companies, and state administration in biomethanation-focused innovation ecosystems.

The main limitation of the research is that it was built on abductive methodology, i.e., conclusions are not confirmed in a positivist sense by testing hypotheses. Because of the specific environment and the integrative, high-level of the analysis, the study might not cover every aspect of biomethanation technology development and competitiveness. For example, future research could identify the competitiveness-increasing potential of biomethanation in future energy systems, or maybe compare it to the potential of chemical methanation. The abductively revealed key topics, however, such as the role of biomethanation in the hydrogen economy or integration with carbon capture technologies, the alignment of technology development, scientific research, and policies might induce other thoughts and future research that facilitates the broad utilization of this innovative technology.

Author Contributions: Conceptualization, A.R.I. and Z.C.; methodology, M.Z. and G.P.; validation, A.R.I. and Z.C., formal analysis, G.P. and M.Z.; investigation, A.R.I., G.P., Z.C. and M.Z.; data curation, M.Z.; writing—original draft preparation, G.P. and M.Z.; writing—review and editing, A.R.I. and

Z.C.; visualization, M.Z.; supervision, A.R.I.; project administration, Z.C. All authors have read and agreed to the published version of the manuscript.

Funding: This research received no external funding.

Acknowledgments: The authors would like to thank Hiventures Zrt./State Fund for Research and Development and Innovation for their investment that enabled this research. This work was prepared with the professional support of the doctoral student scholarship program of the Co-Operative Doctoral Program of the Ministry of Innovation and Technology financed from the National Research, Development and Innovation Fund. Part of this work was performed in the frame of the 2020-3.1.1-ZFR-KVG-2020-00006 project, implemented with the support provided from the National Research, Development and Innovation Fund of Hungary, financed under the 2020-3.1.2-ZFR-KVG funding scheme.

Conflicts of Interest: Two of the authors perform their P2G research at Corvinus University of Budapest, and they have founded the innovative startup company Power-to-Gas Hungary Kft. in order to perform industrial R&D and further develop the technology in pre-commercial and commercial environments.

Abbreviations

ATOM	Abductive theory of method
CNG	Compressed Natural Gas
CCU	Carbon Capture and utilization
EMG-BES	Bioelectrochemical system for electromethanogenesis
EU	European Union
GHG	Greenhouse gas
HSU	Hydrogen storage and utilization
LNG	Liquified natural gas
P2G	Power-to-gas
P2H	Power-to-Hydrogen
P2L	Power-to-liquid
P2M	Power-to-methane
P2X	Power-to-X
R&D&I	Research, development and innovation
SNG	Synthetic natural gas

Appendix A

Table A1. Selected publications for quantitative text analysis.

Author	Year	Title	Keywords
Agneessens et al. [37]	2017	In-situ biogas upgrading with pulse H_2 additions: The relevance of methanogen adaption and inorganic carbon level	Acetate, CO_2 affinity, H_2, Hydrogenotrophic methanogenesis, In situ biogas upgrading
Alitalo et al. [114]	2015	Biocatalytic methanation of hydrogen and carbon dioxide in a fixed bed bioreactor	Hydrogen, Hydrogenotrophic methanogens, Methane production, Methanogens, Power to gas
Amez et al. [115]	2021	Underground methanation, a natural way to transform carbon dioxide into methane	Sustainable energy, Energy storage, Underground hydrogen storage, Green hydrogen, Underground methanization, Methanization
Ancona et al. [116]	2020	Numerical prediction of off-design performance for a Power-to-Gas system coupled with renewables	Power-to-Gas, Co-electrolysis, Methanation, Storage system, Off-design performance, Renewables
Bacariza et al. [117]	2019	Power-to-methane over Ni/zeolites: Influence of the framework type	CO_2 methanation, USY, BEA, ZSM-5, MOR, Nickel
Bailera et al. [118]	2021	Lab-scale experimental tests of power to gas-oxycombustion hybridization: System design and preliminary results	Power-to-Gas, Oxycombustion, Methanation, Lab-scale facility

Table A1. *Cont.*

Author	Year	Title	Keywords
Bareschino et al. [119]	2020	Feasibility analysis of a combined chemical looping combustion and renewable-energy-based methane production system for CO_2 capture and utilization	Thermal power plants, chemical looping combustion, CO_2 capture and utilization, methanation, numerical model
Bareschino et al. [120]	2020	Life cycle assessment and feasibility analysis of a combined chemical looping combustion and power-to-methane system for CO_2 capture and utilization	CO_2 capture and utilization
Bargiacchi [18]	2021	Power-to-Fuel existing plants and pilot projects	Carbon capture, Green ammonia, Green methanol, Electrolysis, Energy storage, Power-to-ammonia, Power-to-fuel, Power-to-methane, Power-to-methanol, substitute natural gas
Bargiacchi et al. [121]	2021	Power to methane	-
Bedoić et al. [122]	2021	Synergy between feedstock gate fee and power-to-gas: An energy and economic analysis of renewable methane production in a biogas plant	Biogas, Food waste, Optimisation, Uncertainty, Renewable gas
Biswas et al. [123]	2020	A Review on Synthesis of Methane as a Pathway for Renewable Energy Storage With a Focus on Solid Oxide Electrolytic Cell-Based Processes	Renewable fuel, Power-to-X, Hydrogen, Methane, Solid oxide electrolyzer
Blanco et al. [124]	2020	Life cycle assessment integration into energy system models: An application for Power-to-Methane in the EU	TIMES, Ecoinvent, Consequential LCA, Environmental impact, Ex-post analysis, Power-to-Gas
Böhm et al. [59]	2020	Projecting cost development for future large-scale power-to-gas implementations by scaling effects	Power-to-gas, Electrolysis, Methanation, Scaling effects, Technological learning
Carrera & Azzaro-Pantel [125]	2021	Bi-objective optimal design of Hydrogen and Methane Supply Chains based on Power-to-Gas systems	Power-to-Gas, Methanation, Hydrogen, MILP, Augmented epsilon constraint, GAMS, Optimization approach
Chellapandi & Prathiviraj [126]	2020	Methanothermobacter thermautotrophicus strain ΔH as a potential microorganism for bioconversion of CO_2 to methane	Methanogenesis, Methane, Methanothermobacter, Biogas, Systems biology, Power-to-Gas
Csedő et al. [6]	2020	Seasonal Energy Storage Potential Assessment of WWTPs with Power-to-Methane Technology	Seasonal energy storage; Power-to-methane; Wastewater treatment plants; Techno-economic assessment
Dedov et al. [127]	2018	Partial Oxidation of Methane to Synthesis Gas	Synthesis gas, Partial oxidation of methane, Neodymium–calcium cobaltate–nickelate
Fózer et al. [128]	2020	Bioenergy with carbon emissions capture and utilisation towards GHG neutrality: Power-to-Gas storage via hydrothermal gasification	Carbon dioxide utilisation; Power-to-Gas; Carbon Neutral; Catalytic Hydrothermal Gasification; VRE storage; LCA
Gantenbein et al. [129]	2021	Flexible application of biogas upgrading membranes for hydrogen recycle in power-to-methane processes	Power-to-Gas, Biogas, Membrane, Upgrading, Flexibility
Ghaib & Ben-Fares [29]	2018	Power-to-Methane: A state-of-the-art review	CO_2 recycling, Demonstration plants, MethanationPower-to-Methane, Water electrolysis
Giglio et al. [130]	2021	Dynamic modelling of methanation reactors during start-up and regulation in intermittent power-to-gas applications	Power-to-Gas; CO_2 methanation; Synthetic natural gas; Reactor design; Dynamic modelling
Gong et al. [131]	2021	Power-to-X: Lighting the Path to a Net-Zero-Emission Future	Electrical energy, Power, Fossil fuels, Electrocatalysts, Solar energy
Guilarte & Azzaro-Pantel [132]	2020	A Methodological Design Framework for Hybrid "Power-to-Methane" and "Power-to-Hydrogen" Supply Chains: application to Occitania Region, France	Power-to-Gas, Hydrogen, Methane, MILP, Gams
Hermesmann et al. [133]	2021	Promising pathways: The geographic and energetic potential of power-to-x technologies based on regeneratively obtained hydrogen	Energy storage, Wind power, Hydrogen, Carbon dioxide, Power-to-Xfuels
Hervy et al. [134]	2021	Power-to-gas: CO_2 methanation in a catalytic fluidized bed reactor at demonstration scale, experimental results and simulation	Power-to-gas, CO_2 valorization, Catalytic methanation, Demonstration reactor, Fluidized bed reactor

Table A1. *Cont.*

Author	Year	Title	Keywords
Hidalgo & Martín-Marroquín [135]	2020	Power-to-methane, coupling CO_2 capture with fuel production: An overview	Biological CO_2 methanation, Chemical CO_2 methanation, Catalityc CO_2 methanation, Carbon capture, Energy storage, Power-to-Gas
Hoffarth et al. [136]	2019	Effect of N_2 on Biological Methanation in a Continuous Stirred-Tank Reactor with Methanothermobacter marburgensis	Biological methanation; CSTR; Methanothermobacter marburgensis; Methane; Carbon dioxide; Dinitrogen; Hydrogen; Power-to-gas
Inkeri et al. [137]	2018	Dynamic one-dimensional model for biological methanation in a stirred tank reactor	Biological methanation, Gas–liquid mass transfer, Power-to-gas, Dynamic model, Stirred tank reactor
Inkeri et al. [138]	2021	Significance of methanation reactor dynamics on the annual efficiency of power-to-gas -system	Power-to-gas; Energy storage; Methanation; Modeling; Wind; Solar
Jentsch et al. [139]	2014	Optimal Use of Power-to-Gas Energy Storage Systems in an 85% Renewable Energy Scenario	Power-to-Gas, Methane, Long-term electricity storage, Economic optimization, Unit commitment
Kassem et al. [140]	2020	Integrating anaerobic digestion, hydrothermal liquefaction, and biomethanation within a power-to-gas framework for dairy waste management and grid decarbonization: a techno-economic assessment	-
Kirchbacher et al. [141]	2018	Process Optimisation of Biogas-Based Power-to-Methane Systems by Simulation	-
Kummer & Imre [142]	2021	Seasonal and Multi-Seasonal Energy Storage by Power-to-Methane Technology	Power-to-Gas; Power-to-Fuel; P2M; P2G; P2F; Biomethanization
Lecker et al. [143]	2017	Biological hydrogen methanation—A review	Biogas, Molecular hydrogen, Carbon dioxide, Power-to-Gas, Energy storage
Leonzio & Zondervan [144]	2020	Analysis and optimization of carbon supply chains integrated to a power to gas process in Italy	CCUS and CCU supply Chain, Mathematical model, Optimization, Reduction of CO_2 emissions
Liao et al. [145]	2020	A Recent Overview of Power-to-Gas Projects	Power-to-Gas, Power-to-Hydrogen, Power-to-Methane
Lin et al. [146]	2020	Geometric synergy of Steam/Carbon dioxide Co-electrolysis and methanation in a tubular solid oxide Electrolysis cell for direct Power-to-Methane	Solid oxide electrolysis cell (SOEC), Steam/carbon dioxide co-electrolysis, Direct power-to-methane, Geometry optimization, Pressurization, Electricity-to-methane efficiency
Liu et al. [147]	2020	The economic and environmental impact of power to hydrogen/power to methane facilities on hybrid power-natural gas energy systems	Power to hydrogen (P2H), Power to methane (P2M), Hydrogen energy, Hybrid power-natural gas energy systems, Renewable energy
Lovato et al. [148]	2017	In-situ biogas upgrading process: Modeling and simulations aspects	Biogas upgrading, Hydrogenotrophic methanogens, Mathematical modeling, Sensitivity analysis
Luo et al. [149]	2018	Synchronous enhancement of H_2O/CO_2 co-electrolysis and methanation for efficient one-step power-to-methane	Solid oxide electrolysis cell, One-step power-to-methane, In-situ thermal coupling, Pressurized
Meylan et al. [150]	2017	Power-to-gas through CO_2 methanation: Assessment of the carbon balance regarding EU directives	Power-to-gas, CO_2-fuels, Carbon balance, Renewable Energy Directive, Carbon capture and utilization, CO_2 valorization
Michailos et al. [151]	2021	A techno-economic assessment of implementing power-to-gas systems based on biomethanation in an operating waste water treatment plant	Biomethanation, Power to gas, Biogas upgrading, CO_2 utilisation, Techno-Economics, Carbon footprint assessment
Momeni et al. [152]	2021	A comprehensive analysis of a power-to-gas energy storage unit utilizing captured carbon dioxide as a raw material in a large-scale power plant	CO_2 utilization, Power-to-gas, Process design, Reaction kinetics, CO_2 methanation, SNG
Monzer et al. [153]	2021	Investigation of the Techno-Economical Feasibility of the Power-to-Methane Process Based on Molten Carbonate Electrolyzer	Molten Carbonate Electrolysis Cell, CO_2, Power-to-gas, Methane synthesis, Economic assessment
Morgenthaler et al. [154]	2020	Site-dependent levelized cost assessment for fully renewable Power-to-Methane systems	Synthetic natural gas, Power-to-Methane, Energy systems modeling, Sector coupling, Carbon capture and utilization (CCU)

Table A1. *Cont.*

Author	Year	Title	Keywords
Mulat et al. [155]	2017	Exogenous addition of H_2 for an in situ biogas upgrading through biological reduction of carbon dioxide into methane	In situ biogas upgrading, H_2 addition, Power to gas, Homo-acetogenesis, Stable isotope, CO_2 reduction
Ortiz et al. [156]	2020	Packed-bed and Microchannel Reactors for the Reactive Capture of CO_2 within Power-to-Methane (P2M) Context: A Comparison	Methanation, Microreactor, Packed-bed reactor, Hot spot formation, Computational Fluid Dynamics
Patterson et al. [157]	2017	Integration of Power to Methane in a waste water treatment plant—A feasibility study	Biomethanation, Power to Gas, Power to Methane, Biogas upgrading, Grid balancing
Pieta et al. [158]	2021	CO_2 Hydrogenation to Methane over Ni-Catalysts: The Effect of Support and Vanadia Promoting	CO_2 hydrogenation; methanation; Ni-catalyst; SMR catalysts; vanadium oxide catalysts
Pintér [5]	2020	The Potential Role of Power-to-Gas Technology Connected to Photovoltaic Power Plants in the Visegrad Countries—A Case Study	Power-to-gas; regulation; Energy storage; Biogas; Biomethane
Pörzse et al. [15]	2021	Disruption Potential Assessment of the Power-to-Methane Technology	Power-to-methane; Disruptive technology; Seasonal energy storage; Decarbonization; Innovation
Sánchez et al. [159]	2021	Optimal design of sustainable power-to-fuels supply chains for seasonal energy storage	Power-to-fuels, Chemical energy storage, Power-to-X, Renewable energy
Savvas, et al. [160]	2018	Methanogenic capacity and robustness of hydrogenotrophic cultures	Hydrogenotrophic methanogenesis, Biofilm, Power to gas, Energy storage
Schlautmann et al. [161]	2021	Renewable Power-to-Gas: A Technical and Economic Evaluation of Three Demo Sites Within the STORE&GO Project	Demo sites, Dynamic operation, Efficiency, Future cost development, Investment costs, Power-to-Gas, Production costs
Sinóros-Szabó et al. [162]	2018	Biomethane production monitoring and data analysis based on the practical operation experiences of an innovative power-to-gas benchscale prototype	Biomethane production, Power-to-gas, Prototype, Monitoring and analysis
Stangeland et al. [163]	2017	CO_2 methanation: the effect of catalysts and reaction conditions	Sabatier reaction, CO_2 methanation, energy storage, biogas upgrading, reaction conditions, nickel catalyst
Straka [164]	2021	A comprehensive study of Power-to-Gas technology: Technical implementations overview, economic assessments, methanation plant as auxiliary operation of lignite-fired power station	Power-to-Gas, Energy storage, Electrolysis, Methanation, CO_2 source
Vo et al. [165]	2018	Can power to methane systems be sustainable and can they improve the carbon intensity of renewable methane when used to upgrade biogas produced from grass and slurry?	Life cycle assessment, Sustainability criteria, Advanced biofuels, Power to gas, Biological methanation, Co-digestion
Wang et al. [166]	2018	Optimal design of solid-oxide electrolyzer based power-to-methane systems	Energy storage, Power-to-gas, Power-to-methane, Solid-oxide electrolyzer, Co-electrolysis, CO_2 utilization
Wang et al. [167]	2020	Reversible solid-oxide cell stack based power-to-x-to-power systems: Comparison of thermodynamic performance	Electrical storage, Power-to-x, Reversible solid-oxide cell, Ammonia, Methanol, Sector coupling
Welch et al. [168]	2021	Comparative Technoeconomic Analysis of Renewable Generation of Methane Using Sunlight, Water, and Carbon Dioxide	Atmospheric chemistry, Hydrocarbons, Membranes, Electrical energy, Electrolysis
Xie et al. [169]	2020	Optimization on Combined Cooling, Heat and Power Microgrid System with Power-to-gas Devices	Combined cooling, heat and power, Microgrid, Power-to-grid, Hydrogen natural gas blends
Zoss et al. [170]	2016	Modeling a power-to-renewable methane system for an assessment of power grid balancing options in the Baltic States' region	Excess power, Methanation, Power-to-gas, Power-to-methane, Renewable methane, Stochastic energy

References

1. Lund, H.; Østergaard, P.A.; Connolly, D.; Ridjan, I.; Mathiesen, B.V.; Hvelplund, F.; Thellufsen, J.Z.; Sorknæs, P. Energy Storage and Smart Energy Systems. *Int. J. Sustain. Energy Plan. Manag.* **2016**, *11*, 3–14. [CrossRef]
2. Ahmed, A.M.; Kondor, L.; Imre, A.R. Thermodynamic Efficiency Maximum of Simple Organic Rankine Cycles. *Energies* **2021**, *14*, 307. [CrossRef]
3. Pintér, G.; Zsiborács, H.; Baranyai, N.H.; Vincze, A.; Birkner, Z. The Economic and Geographical Aspects of the Status of Small-Scale Photovoltaic Systems in Hungary—A Case Study. *Energies* **2020**, *13*, 3489. [CrossRef]
4. Berényi, L.; Birkner, Z.; Deutsch, N. A Multidimensional Evaluation of Renewable and Nuclear Energy among Higher Education Students. *Sustainability* **2020**, *12*, 1449. [CrossRef]
5. Pintér, G. The Potential Role of Power-to-Gas Technology Connected to Photovoltaic Power Plants in the Visegrad Countries—A Case Study. *Energies* **2020**, *13*, 6408. [CrossRef]
6. Csedő, Z.; Sinóros-Szabó, B.; Zavarkó, M. Seasonal Energy Storage Potential Assessment of WWTPs with Power-to-Methane Technology. *Energies* **2020**, *13*, 4973. [CrossRef]
7. Csedő, Z.; Zavarkó, M. The role of inter-organizational innovation networks as change drivers in commercialization of disruptive technologies: The case of power-to-gas. *Int. J. Sustain. Energy Plan. Manag.* **2020**, *28*, 53–70. [CrossRef]
8. Strategieplatform Power to Gas. Audi e-Gas Projekt. 2021. Available online: https://www.powertogas.info/projektkarte/audi-e-gas-projekt/ (accessed on 8 March 2021).
9. Electrochaea. Power-to-Gas via Biological Catalysis (P2G-Biocat); Project Final Report. 2017. Available online: https://energiforskning.dk/sites/energiforskning.dk/files/slutrapporter/12164_final_report_p2g_biocat.pdf (accessed on 8 March 2021).
10. Guilera, J.; Morante, J.R.; Andreu, T. Economic viability of SNG production from power and CO_2. *Energy Convers. Manag.* **2018**, *162*, 218–224. [CrossRef]
11. Peters, R.; Baltruweit, M.; Grube, T.; Samsun, R.C.; Stolten, D. A techno economic analysis of the power to gas route. *J. CO_2 Util.* **2019**, *34*, 616–634. [CrossRef]
12. Schiebahn, S.; Grube, T.; Robinius, M.; Tietze, V.; Kumar, B.; Stolten, D. Power to gas: Technological overview, systems analysis and economic assessment for a case study in Germany. *Int. J. Hydrogen Energy* **2015**, *40*, 4285–4294. [CrossRef]
13. Götz, M.; Lefebvre, J.; Mörs, F.; Koch, A.M.; Graf, F.; Bajohr, S.; Reimert, R.; Kolb, T. Renewable Power-to-Gas: A technological and economic review. *Renew. Energy* **2016**, *85*, 1371–1390. [CrossRef]
14. Blanco, H.; Faaij, A. A review at the role of storage in energy systems with a focus on Power to Gas and long-term storage. *Renew. Sustain. Energy Rev.* **2018**, *81*, 1049–1086. [CrossRef]
15. Pörzse, G.; Csedő, Z.; Zavarkó, M. Disruption Potential Assessment of the Power-to-Methane Technology. *Energies* **2021**, *14*, 2297. [CrossRef]
16. Preston, N.; Maroufmashat, A.; Riaz, H.; Barbouti, S.; Mukherjee, U.; Tang, P.; Wang, J.; Haghi, E.; Elkamel, A.; Fowler, M. How can the integration of renewable energy and power-to-gas benefit industrial facilities? From techno-economic, policy, and environmental assessment. *Int. J. Hydrogen Energy* **2020**, *45*, 26559–26573. [CrossRef]
17. Bailera, M.; Lisbona, P.; Romeo, L.M.; Espatolero, S. Power to Gas projects review: Lab, pilot and demo plants for storing renewable energy and CO_2. *Renew. Sustain. Energy Rev.* **2017**, *69*, 292–312. [CrossRef]
18. Bargiacchi, E. Power-to-Fuel existing plants and pilot projects. In *Power to Fuel*; Academic Press: Cambridge, MA, USA, 2021; pp. 211–237.
19. Thema, M.; Bauer, F.; Sterner, M. Power-to-Gas: Electrolysis and methanation status review. *Renew. Sustain. Energy Rev.* **2019**, *112*, 775–787. [CrossRef]
20. Chehade, Z.; Mansilla, C.; Lucchese, P.; Hilliard, S.; Proost, J. Review and analysis of demonstration projects on power-to-X pathways in the world. *Int. J. Hydrogen Energy* **2019**, *44*, 27637–27655. [CrossRef]
21. Brunner, C.; Michaelis, J.; Möst, D. Competitiveness of Different Operational Concepts for Power-to-Gas in Future Energy Systems. *Z. Energ.* **2015**, *39*, 275–293. [CrossRef]
22. Ameli, H.; Qadrdan, M.; Strbac, G. Techno-economic assessment of battery storage and Power-to-Gas: A whole-system approach. *Energy Procedia* **2017**, *142*, 841–848. [CrossRef]
23. Collet, P.; Flottes, E.; Favre, A.; Raynal, L.; Pierre, H.; Capela, S.; Peregrina, C. Techno-economic and Life Cycle Assessment of methane production via biogas upgrading and power to gas technology. *Appl. Energy* **2017**, *192*, 282–295. [CrossRef]
24. Fagerberg, J. Technology and competitiveness. *Oxf. Rev. Econ. Policy* **1996**, *12*, 39–51. [CrossRef]
25. European Commission. *The European Green Deal*; European Commission: Brussels, Belgium, 2019.
26. Brydon-Miller, M.; Coghlan, D. The big picture: Implications and imperatives for the action research community from the SAGE Encyclopedia of Action Research. *Action Res.* **2014**, *12*, 224–233. [CrossRef]
27. Ferry, J.G. Enzymology of one-carbon metabolism in methanogenic pathways. *FEMS Microbiol. Rev.* **1998**, *23*, 13–38. [CrossRef] [PubMed]
28. Kim, S.; Yang, Y.; Lippi, R.; Choi, H.; Kim, S.; Chun, D.; Im, H.; Lee, S.; Yoo, J. Low-Rank Coal Supported Ni Catalysts for CO_2 Methanation. *Energies* **2021**, *14*, 2040. [CrossRef]
29. Ghaib, K.; Ben-Fares, F.-Z. Power-to-Methane: A state-of-the-art review. *Renew. Sustain. Energy Rev.* **2018**, *81*, 433–446. [CrossRef]
30. Leeuwen, C. *Report on the Costs Involved with P2G Technologies and Their Potentials across the EU*; STORE&GO: Bonn, Germany, 2018.

31. Electrochaea. How the Technology Works. 2019. Available online: http://www.electrochaea.com/technology/ (accessed on 18 March 2019).
32. Frontera, P.; Macario, A.; Ferraro, M.; Antonucci, P. Supported Catalysts for CO_2 Methanation: A Review. *Catalysts* **2017**, *7*, 59. [CrossRef]
33. Markthaler, S.; Plankenbühler, T.; Weidlich, T.; Neubert, M.; Karl, J. Numerical simulation of trickle bed reactors for biological methanation. *Chem. Eng. Sci.* **2020**, *226*, 115847. [CrossRef]
34. Gutiérrez-Martín, F.; Rodríguez-Antón, L.; Legrand, M. Renewable power-to-gas by direct catalytic methanation of biogas. *Renew. Energy* **2020**, *162*, 948–959. [CrossRef]
35. Shen, L.; Xu, J.; Zhu, M.; Han, Y.-F. Essential Role of the Support for Nickel-Based CO_2 Methanation Catalysts. *ACS Catal.* **2020**, *10*, 14581–14591. [CrossRef]
36. Ács, N.; Szuhaj, M.; Wirth, R.; Bagi, Z.; Maróti, G.; Rákhely, G.; Kovács, K.L. Microbial Community Rearrangements in Power-to-Biomethane Reactors Employing Mesophilic Biogas Digestate. *Front. Energy Res.* **2019**, *7*, 1–15. [CrossRef]
37. Agneessens, L.M.; Ottosen, L.D.M.; Voigt, N.V.; Nielsen, J.L.; de Jonge, N.; Fischer, C.H.; Kofoed, M.V.W. In-situ biogas upgrading with pulse H 2 additions: The relevance of methanogen adaption and inorganic carbon level. *Bioresour. Technol.* **2017**, *233*, 256–263. [CrossRef] [PubMed]
38. Ceballos-Escalera, A.; Molognoni, D.; Bosch-Jimenez, P.; Shahparasti, M.; Bouchakour, S.; Luna, A.; Guisasola, A.; Borràs, E.; Della Pirriera, M. Bioelectrochemical systems for energy storage: A scaled-up power-to-gas approach. *Appl. Energy* **2020**, *260*, 114138. [CrossRef]
39. Jambor, A.; Babu, S. Competitiveness: Definitions, Theories and Measurement. In *Competitiveness of Global Agriculture*; Springer: Cham, Switzerland, 2016; pp. 25–45. [CrossRef]
40. Cantwell, J. Innovation and Competitiveness. In *The Oxford Handbook of Innovation*; Fagerber, J., Mowery, D.C., Nelson, R.R., Eds.; Oxford University Press: New York, NY, USA, 2005; pp. 543–567.
41. Tidd, J.; Bessant, J.; Pavitt, K. *Managing Innovation*; Wiley: Chichester, UK, 1997.
42. Teece, D.J. Profiting from technological innovation: Implications for integration, collaboration, licensing and public policy. *Res. Policy* **1986**, *15*, 285–305. [CrossRef]
43. Mascarenhas, C.; Ferreira, J.; Marques, C. University-industry cooperation: A systematic literature review and research agenda. *Sci. Public Policy* **2018**, *45*, 708–718. [CrossRef]
44. Glaser, B.; Strauss, A. *The Discovery of Grounded Theory: Strategies for Qualitative Research*; Aldine: Chicago, IL, USA, 1967.
45. Burawoy, M. The Extended Case Method. *Sociol. Theory* **1998**, *16*, 4–33. [CrossRef]
46. Haig, B.D. An Abductive Theory of Scientific Method. In *Method Matters in Psychology*; Studies in Applied Philosophy, Epistemology and Rational Ethics; Springer: Cham, Switzerland, 2018; Volume 45, pp. 35–64. [CrossRef]
47. European Commission. *A Clean Planet for All—A European Strategic Long-Term Vision for a Prosperous, Modern, Competitive and Climate Neutral Economy—Communication from the Commission to the European Parliament*; European Council: Brussels, Belgium, 2018.
48. European Commission. *The European Green Deal—Communication from the Commission to the European Parliament*; European Council: Brussels, Belgium, 2019.
49. European Commission. *The European Green Deal—Annex to the Communication from the Commission to the European Parliament*; European Council: Brussels, Belgium, 2019.
50. European Parliament. *The European Green Deal—European Parliament Resolution of 15 January 2020 on the European Green Deal (2019/2956(RSP)), P9_TA(2020)0005*; European Parliament: Brussels, Belgium, 2020.
51. Zengul, F.D.; Zengul, A.G.; Mugavero, M.J.; Oner, N.; Ozaydin, B.; Delen, D.; Willig, J.H.; Kennedy, K.C.; Cimino, J. A critical analysis of COVID-19 research literature: Text mining approach. *Intell. Med.* **2021**, *5*, 100036. [CrossRef]
52. Kimes, P.; Liu, Y.; Hayes, D.N.; Marron, J.S. Statistical significance for hierarchical clustering. *Biometrics* **2017**, *73*, 811–821. [CrossRef]
53. Innovation Land Lab. Projekte BioCO$_2$nvert. 2021. Available online: https://innovation-landlab.de/projekte/bioco2nvert/ (accessed on 14 July 2021).
54. Technical University of Ostwestfalen-Lippe. BioCO$_2$nvert. 2021. Available online: https://www.th-owl.de/ilt-nrw/projekte/bioco2nvert/ (accessed on 14 July 2021).
55. Burgundy-Franche-Comté Regional Council. HyCAUNAIS. 2021. Available online: https://www.europe-bfc.eu/beneficiaire/hycaunais/ (accessed on 14 July 2021).
56. Ademe. HyCAUNAIS V2. 2019. Available online: https://librairie.ademe.fr/recherche-et-innovation/500-hycaunais-v2.html (accessed on 14 July 2021).
57. Zhaw. CarbonATE—Microbiological Methanation. 2021. Available online: https://www.zhaw.ch/en/research/research-database/project-detailview/projektid/3023/ (accessed on 14 July 2021).
58. Energy Innovation Austria. CarbonATE. 2021. Available online: https://www.energy-innovation-austria.at/article/carbonate-2/?lang=en (accessed on 14 July 2021).
59. Böhm, H.; Zauner, A.; Rosenfeld, D.C.; Tichler, R. Projecting cost development for future large-scale power-to-gas implementations by scaling effects. *Appl. Energy* **2020**, *264*, 114780. [CrossRef]

60. Blanco, H.; Nijs, W.; Ruf, J.; Faaij, A. Potential of Power-to-Methane in the EU energy transition to a low carbon system using cost optimization. *Appl. Energy* **2018**, *232*, 323–340. [CrossRef]
61. German Energy Agency GmbH (dena). Viessmann Power-to-Gas in the Eucolino in Schwandorf. 2021. Available online: https://www.powertogas.info/projektkarte/viessmann-power-to-gas-im-eucolino-in-schwandorf/ (accessed on 19 July 2021).
62. Electrochaea. Electrochaea Commissions World's Largest Power-To-Gas Demonstration Project. 13 August 2013. Available online: https://www.electrochaea.com/electrochaea-commissions-worlds-largest-power-to-gas-demonstration-project-based-on-biological-methanation/ (accessed on 16 July 2021).
63. Iskov, H.; Rasmussen, N.B. *Global Screening of Projects and Technologies for Power-to-Gas and Bio-SNG*; Danish Gas Technology Centre: Hørsholm, Denmark, 2013.
64. Panagiotis, T.; Angelidaki, I. *Final Project Report-SYMBIO (12-132654)*; Technical University of Denmark: Lyngby, Denmark, 2020.
65. Attero. Our Locations. 2021. Available online: https://www.attero.nl/nl/onze-locaties/wijster/ (accessed on 16 July 2021).
66. Heller, T. BioPower2Gas—Power-to-Gas with Biological Methanation. In Proceedings of the Biomass for Swiss Energy Future Conference, Brugg, Switzerland, 7 September 2016.
67. IEA Bioenergy. BioPower2Gas in Germany. 2018. Available online: https://www.ieabioenergy.com/wp-content/uploads/2018/02/2-BioPower2Gas_DE_Final.pdf (accessed on 16 July 2021).
68. Strategieplattform Power to Gas. Trickle-Bed Reactor for Biological Methanation in the Large-Scale Technical Center of GICON. 2021. Available online: https://www.powertogas.info/projektkarte/rieselbettreaktor-fuer-die-biologische-methanisierung-im-grosstechnikum-der-gicon/ (accessed on 16 July 2021).
69. GICON®-Großmann Ingenieur Consult GmbH, Biogastechnikum. 2021. Available online: https://www.gicon.de/leistungen-gicon-consult/forschung/biogastechnikum (accessed on 19 July 2021).
70. PFI Germany. Biorefinery at Pirmasens Winzeln Energy Park. 2021. Available online: https://www.pfi-germany.de/en/research/biorefinery-at-pirmasens-winzeln-energy-park/ (accessed on 16 July 2021).
71. Energie-Atlas Bayern. Mikrobielle Methanisierung Speicherung Elektrischer Überschussenergie durch Methansierung von Klärgas. 2019. Available online: https://www.energieatlas.bayern.de/thema_sonne/photovoltaik/praxisbeispiele/details,704.html (accessed on 19 July 2021).
72. Viebahn, P.; Zelt, O.; Fischedick, M.; Wietschel, M.; Hirzel, S.; Horst, J. *Technologien für die Energiewende Technologiebericht—Band 2*; Wuppertal Institut für Klima, Umwelt Energie GmbH: Wuppertal, Germany, 2018.
73. Strategieplattform Power to Gas. Power to Gas Biogasbooster. 2021. Available online: https://www.powertogas.info/projektkarte/power-to-gas-biogasbooster/ (accessed on 16 July 2021).
74. Hannula, I.; Hakkarainen, E. *Integrated Bioenergy Hybrids: Flexible Renewable Energy Solutions*; IEA Bioenergy: Paris, France, 2017.
75. Electrochaea. Energy Storage Leaders Launch Commercial Scale Power-To-Gas Project. 2014. Available online: https://www.electrochaea.com/energy-storage-leaders-launch-commercial-scale-power-to-gas-project-using-highly-innovative-technology/ (accessed on 16 July 2021).
76. Lardon, L.; Thorberg, D.; Krosgaard, L. *WP3—Biogas Valorization and Efficient Energy Management—D 3.2: Technical and Economic Analysis of Biological Methanation*; Powerstep: West Chester, OH, USA, 2018.
77. MicroPyros. Reaktorbau. 2021. Available online: https://www.micropyros.de/leistungen-und-service/reaktorbau/ (accessed on 19 July 2021).
78. STORE&GO. The STORE&GO Demonstration Site at Solothurn, Switzerland. 2021. Available online: https://www.storeandgo.info/demonstration-sites/switzerland/ (accessed on 16 July 2021).
79. FAU. BMWi Projekt ORBIT—Bmwi Optimization of a Trickle-Bed-Bioreactor for Dynamic Microbial Biosynthesis of Methane with Archaea in Power-To-Gas Plants. 2021. Available online: https://www.evt.tf.fau.eu/research/schwerpunkte/2nd-generation-fuels/orbit/#collapse_3 (accessed on 16 July 2021).
80. Ademe. HyCAUNAIS V2. 2019. Available online: https://www.ademe.fr/sites/default/files/assets/documents/hycaunais_v2_vf.pdf (accessed on 16 July 2021).
81. Viessmann. Grünes Licht für Erste Industrielle Power-to-Gas-Anlage. 2019. Available online: https://www.viessmann.family/de/newsroom/unternehmen/gruenes-licht-fuer-erste-industrielle-power-to-gas-anlage (accessed on 16 July 2021).
82. Renewable Carbon, German Government Committed to the Widespread Use of Power-to-Methane Technology. 2020. Available online: https://renewable-carbon.eu/news/german-government-committed-to-the-widespread-use-of-power-to-methane-technology/ (accessed on 16 July 2021).
83. Electrochaea. Electrochaea Realisiert Power-to-Gas- Anlage für ein Nachhaltiges Pfaffenhofen. 2017. Available online: http://www.electrochaea.com/wp-content/uploads/2017/11/20171113_PM-Electrochaea_PtoG_fuer_Pfaffenhofen_DE_FIN.pdf (accessed on 16 July 2021).
84. Power-to-Gas Hungary Kft. About. 2021. Available online: https://p2g.hu (accessed on 16 July 2021).
85. Abe, J.; Popoola, A.; Ajenifuja, E.; Popoola, O. Hydrogen energy, economy and storage: Review and recommendation. *Int. J. Hydrogen Energy* **2019**, *44*, 15072–15086. [CrossRef]
86. STORE&GO Project. Results and Publications of the Project. 2021. Available online: https://www.storeandgo.info (accessed on 1 July 2021).
87. ITM. *National Energy Strategy 2030*; Ministry of National Development: Budapest, Hungary, 2020.

88. Haeseldonckx, D.; Dhaeseleer, W. The use of the natural-gas pipeline infrastructure for hydrogen transport in a changing market structure. *Int. J. Hydrogen Energy* **2007**, *32*, 1381–1386. [CrossRef]
89. Messaoudani, Z.L.; Rigas, F.; Hamid, M.D.B.; Hassan, C.R.C. Hazards, safety and knowledge gaps on hydrogen transmission via natural gas grid: A critical review. *Int. J. Hydrogen Energy* **2016**, *41*, 17511–17525. [CrossRef]
90. Bompard, E.; Bensaid, S.; Chicco, G.; Mazza, A. *Report on the Model of the Power System with P2G*; STORE&GO: Bonn, Germany, 2018.
91. Jepma, C.; van Leeuwen, C.; Hulshof, D. *Exploring the Future for Green Gases*; STORE&GO: Bonn, Germany, 2017.
92. Van der Welle, A.J.; De Nooij, M.; Mozaffarian, M. *Full Socio-Economic Costs and Benefits of Energy Mix Diversification and the Role of Power-To-Gas in This Regard*; STORE&GO: Bonn, Germany, 2018.
93. Carnegie Mellon University. *IECM Technical Documentation: Amine-based Post-Combustion CO_2 Capture*; IECM: Pittsburgh, PA, USA, 2018.
94. Stanger, R.; Wall, T.; Spörl, R.; Paneru, M.; Grathwohl, S.; Weidmann, M.; Scheffknecht, G.; McDonald, D.; Myöhänen, K.; Ritvanen, J.; et al. Oxyfuel combustion for CO_2 capture in power plants. *Int. J. Greenh. Gas Control.* **2015**, *40*, 55–125. [CrossRef]
95. Bailera, M.; Lisbona, P.; Peña, B.; Romeo, L.M. Integration of Amine Scrubbing and Power to Gas. In *Energy Storage*; Springer: Cham, Switzerland, 2020; pp. 109–135. [CrossRef]
96. Bailera, M.; Espatolero, S.; Lisbona, P.; Romeo, L.M. Power to gas-electrochemical industry hybrid systems: A case study. *Appl. Energy* **2017**, *202*, 435–446. [CrossRef]
97. Teece, D.J. Explicating dynamic capabilities: The nature and microfoundations of (sustainable) enterprise performance. *Strat. Manag. J.* **2007**, *28*, 1319–1350. [CrossRef]
98. Oosterkamp, P. *Full CBA Analysis of Power-to-Gas in the Context of Various Reference Scenarios*; STORE&GO: Bonn, Germany, 2018.
99. Blanco, H. *Report on Full CBA Based on the Relevant Environmental Impact Data*; STORE&GO: Bonn, Germany, 2018.
100. Dumas, C.; Ottosen, L.D.M.; Escudié, R.; Jensen, P. Editorial: Biological Methanation or (Bio/Syn)-Gas Upgrading. *Front. Energy Res.* **2020**, *8*, 30. [CrossRef]
101. Azarova, V.; Cohen, J.; Friedl, C.; Reichl, J. *Report on Social and Public Acceptance Determinants in Selected EU-Countries*; STORE&GO: Bonn, Germany, 2019.
102. Boyer, J. Toward an Evolutionary and Sustainability Perspective of the Innovation Ecosystem: Revisiting the Panarchy Model. *Sustainability* **2020**, *12*, 3232. [CrossRef]
103. Porter, M.E. *Competitive Advantage of Nations*; The Free Press: New York, NY, USA, 1990.
104. Clark, J.; Guy, K. Innovation and competitiveness: A review. *Technol. Anal. Strat. Manag.* **1998**, *10*, 363–395. [CrossRef]
105. Schwab, K.; Zahidi, S. *The Global Competitiveness Report Special Edition 2020 How Countries are Performing on the Road to Recovery*; World Economic Forum: Cologny/Geneva, Switzerland, 2020; ISBN 978-2-940631-17-9.
106. Möbius, P.; Althammer, W. Sustainable competitiveness: A spatial econometric analysis of European regions. *J. Environ. Plan. Manag.* **2020**, *63*, 453–480. [CrossRef]
107. Kelly, L.M.; Cordeiro, M. Three principles of pragmatism for research on organizational processes. *Methodol. Innov.* **2020**, *13*. [CrossRef]
108. Kirchherr, J.; Reike, D.; Hekkert, M. Conceptualizing the circular economy: An analysis of 114 definitions. *Resour. Conserv. Recycl.* **2017**, *127*, 221–232. [CrossRef]
109. Baena-Moreno, F.M.; Zhang, Z.; Zhang, X.; Reina, T. Profitability analysis of a novel configuration to synergize biogas upgrading and Power-to-Gas. *Energy Convers. Manag.* **2020**, *224*, 113369. [CrossRef]
110. Eggemann, L.; Escobar, N.; Peters, R.; Burauel, P.; Stolten, D. Life cycle assessment of a small-scale methanol production system: A Power-to-Fuel strategy for biogas plants. *J. Clean. Prod.* **2020**, *271*, 122476. [CrossRef]
111. D'Adamo, I.; Falcone, P.M.; Huisingh, D.; Morone, P. A circular economy model based on biomethane: What are the opportunities for the municipality of Rome and beyond? *Renew. Energy* **2021**, *163*, 1660–1672. [CrossRef]
112. Wall, D.; McDonagh, S.; Murphy, J.D. Cascading biomethane energy systems for sustainable green gas production in a circular economy. *Bioresour. Technol.* **2017**, *243*, 1207–1215. [CrossRef]
113. Morone, P.; Caferra, R.; D'Adamo, I.; Falcone, P.M.; Imbert, E.; Morone, A. Consumer willingness to pay for bio-based products: Do certifications matter? *Int. J. Prod. Econ.* **2021**, *240*, 108248. [CrossRef]
114. Alitalo, A.; Niskanen, M.; Aura, E. Biocatalytic methanation of hydrogen and carbon dioxide in a fixed bed bioreactor. *Bioresour. Technol.* **2015**, *196*, 600–605. [CrossRef] [PubMed]
115. Amez, I.; Gonzalez, S.; Sanchez-Martin, L.; Ortega, M.F.; Llamas, B. Underground methanation, a natural way to transform carbon dioxide into methane. In *Causes, Effects and Solutions for Global Warming*; Elsevier: Amsterdam, The Netherlands, 2021; pp. 81–106. [CrossRef]
116. Ancona, M.; Bianchi, M.; Branchini, L.; Catena, F.; De Pascale, A.; Melino, F.; Peretto, A. Numerical prediction of off-design performance for a Power-to-Gas system coupled with renewables. *Energy Convers. Manag.* **2020**, *210*, 112702. [CrossRef]
117. Bacariza, M.; Maleval, M.; Graça, I.; Lopes, J.; Henriques, C. Power-to-methane over Ni/zeolites: Influence of the framework type. *Microporous Mesoporous Mater.* **2019**, *274*, 102–112. [CrossRef]
118. Bailera, M.; Peña, B.; Lisbona, P.; Marín, J.; Romeo, L.M. Lab-scale experimental tests of power to gas-oxycombustion hybridization: System design and preliminary results. *Energy* **2021**, *226*, 120375. [CrossRef]

119. Bareschino, P.; Mancusi, E.; Pepe, F.; Urciuolo, M.; Coppola, A. Feasibility analysis of a combined chemical looping combustion and renewable-energy-based methane production system for CO_2 capture and utilization. *Therm. Sci.* **2020**, *24*, 3613–3624. [CrossRef]
120. Bareschino, P.; Mancusi, E.; Urciuolo, M.; Paulillo, A.; Chirone, R.; Pepe, F. Life cycle assessment and feasibility analysis of a combined chemical looping combustion and power-to-methane system for CO_2 capture and utilization. *Renew. Sustain. Energy Rev.* **2020**, *130*, 109962. [CrossRef]
121. Bargiacchi, E.; Candelaresi, D.; Spazzafumo, G. 4—Power to methane. In *Power to Fuel*; Elsevier: Amsterdam, The Netherlands, 2021; pp. 75–101. ISBN 978-0-12-822813-5.
122. Bedoić, R.; Dorotić, H.; Schneider, D.R.; Čuček, L.; Ćosić, B.; Pukšec, T.; Duić, N. Synergy between feedstock gate fee and power-to-gas: An energy and economic analysis of renewable methane production in a biogas plant. *Renew. Energy* **2021**, *173*, 12–23. [CrossRef]
123. Biswas, S.; Kulkarni, A.P.; Giddey, S.; Bhattacharya, S. A Review on Synthesis of Methane as a Pathway for Renewable Energy Storage with a Focus on Solid Oxide Electrolytic Cell-Based Processes. *Front. Energy Res.* **2020**, *8*, 570112. [CrossRef]
124. Blanco, H.; Codina, V.; Laurent, A.; Nijs, W.; Maréchal, F.; Faaij, A. Life cycle assessment integration into energy system models: An application for Power-to-Methane in the EU. *Appl. Energy* **2020**, *259*, 114160. [CrossRef]
125. Carrera, E.; Azzaro-Pantel, C. Bi-objective optimal design of Hydrogen and Methane Supply Chains based on Power-to-Gas systems. *Chem. Eng. Sci.* **2021**, *246*, 116861. [CrossRef]
126. Chellapandi, P.; Prathiviraj, R. Methanothermobacter thermautotrophicus strain ΔH as a potential microorganism for bioconversion of CO_2 to methane. *J. CO_2 Util.* **2020**, *40*, 101210. [CrossRef]
127. Dedov, A.G.; Shlyakhtin, O.A.; Loktev, A.S.; Mazo, G.N.; Malyshev, S.A.; Tyumenova, S.I.; Baranchikov, A.E.; Moiseev, I.I. Partial oxidation of methane to synthesis gas: Novel catalysts based on neodymium–calcium cobaltate–nickelate complex oxides. *Pet. Chem.* **2018**, *58*, 43–47. [CrossRef]
128. Fózer, D.; Volanti, M.; Passarini, F.; Varbanov, P.S.; Klemeš, J.J.; Mizsey, P. Bioenergy with carbon emissions capture and utilisation towards GHG neutrality: Power-to-Gas storage via hydrothermal gasification. *Appl. Energy* **2020**, *280*, 115923. [CrossRef]
129. Gantenbein, A.; Witte, J.; Biollaz, S.M.; Kröcher, O.; Schildhauer, T.J. Flexible application of biogas upgrading membranes for hydrogen recycle in power-to-methane processes. *Chem. Eng. Sci.* **2021**, *229*, 116012. [CrossRef]
130. Giglio, E.; Pirone, R.; Bensaid, S. Dynamic modelling of methanation reactors during start-up and regulation in intermittent power-to-gas applications. *Renew. Energy* **2021**, *170*, 1040–1051. [CrossRef]
131. Gong, J.; English, N.J.; Pant, D.; Patzke, G.R.; Protti, S.; Zhang, T. Power-to-X: Lighting the Path to a Net-Zero-Emission Future. *ACS Sustain. Chem. Eng.* **2021**, *9*, 7179–7181. [CrossRef]
132. Guilarte, E.C.; Azzaro-Pantel, C. A Methodological Design Framework for Hybrid "Power-to-Methane" and "Power-to-Hydrogen" Supply Chains: Application to Occitania Region, France. *Comput. Aided Chem. Eng.* **2020**, *48*, 679–684. [CrossRef]
133. Hermesmann, M.; Grübel, K.; Scherotzki, L.; Müller, T. Promising pathways: The geographic and energetic potential of power-to-x technologies based on regeneratively obtained hydrogen. *Renew. Sustain. Energy Rev.* **2020**, *138*, 110644. [CrossRef]
134. Hervy, M.; Maistrello, J.; Brito, L.; Rizand, M.; Basset, E.; Kara, Y.; Maheut, M. Power-to-gas: CO_2 methanation in a catalytic fluidized bed reactor at demonstration scale, experimental results and simulation. *J. CO_2 Util.* **2021**, *50*, 101610. [CrossRef]
135. Hidalgo, D.; Martín-Marroquín, J. Power-to-methane, coupling CO_2 capture with fuel production: An overview. *Renew. Sustain. Energy Rev.* **2020**, *132*, 110057. [CrossRef]
136. Hoffarth, M.P.; Broeker, T.; Schneider, J. Effect of N2 on Biological Methanation in a Continuous Stirred-Tank Reactor with Methanothermobacter marburgensis. *Fermentation* **2019**, *5*, 56. [CrossRef]
137. Inkeri, E.; Tynjälä, T.; Laari, A.; Hyppänen, T. Dynamic one-dimensional model for biological methanation in a stirred tank reactor. *Appl. Energy* **2018**, *209*, 95–107. [CrossRef]
138. Inkeri, E.; Tynjälä, T.; Karjunen, H. Significance of methanation reactor dynamics on the annual efficiency of power-to-gas system. *Renew. Energy* **2021**, *163*, 1113–1126. [CrossRef]
139. Jentsch, M.; Trost, T.; Sterner, M. Optimal Use of Power-to-Gas Energy Storage Systems in an 85% Renewable Energy Scenario. *Energy Procedia* **2014**, *46*, 254–261. [CrossRef]
140. Kassem, N.; Hockey, J.; Lopez, C.; Lardon, L.; Angenent, L.T.; Tester, J.W. Integrating anaerobic digestion, hydrothermal liquefaction, and biomethanation within a power-to-gas framework for dairy waste management and grid decarbonization: A techno-economic assessment. *Sustain. Energy Fuels* **2020**, *4*, 4644–4661. [CrossRef]
141. Kirchbacher, F.; Miltner, M.; Wukovits, W.; Friedl, A.; Harasek, M. Process Optimisation of Biogas-Based Power-to-Methane Systems by Simulation. *Chem. Eng. Trans.* **2018**, *70*, 907–912. [CrossRef]
142. Kummer, K.; Imre, A. Seasonal and Multi-Seasonal Energy Storage by Power-to-Methane Technology. *Energies* **2021**, *14*, 3265. [CrossRef]
143. Lecker, B.; Illi, L.; Lemmer, A.; Oechsner, H. Biological hydrogen methanation—A review. *Bioresour. Technol.* **2017**, *245*, 1220–1228. [CrossRef]
144. Leonzio, G.; Zondervan, E. Analysis and optimization of carbon supply chains integrated to a power to gas process in Italy. *J. Clean. Prod.* **2020**, *269*, 122172. [CrossRef]
145. Liao, M.; Liu, C.; Qing, Z. A Recent Overview of Power-to-Gas Projects. In *4th Conference on Energy Internet and Energy System Integration (EI2) 2020*; IEEE: Piscataway, NJ, USA, 2020; pp. 2282–2286. [CrossRef]

146. Lin, L.; Chen, S.; Quan, J.; Liao, S.; Luo, Y.; Chen, C.; Au, C.-T.; Shi, Y.; Jiang, L. Geometric synergy of Steam/Carbon dioxide Co-electrolysis and methanation in a tubular solid oxide Electrolysis cell for direct Power-to-Methane. *Energy Convers. Manag.* **2020**, *208*, 112570. [CrossRef]
147. Liu, J.; Sun, W.; Harrison, G.P. The economic and environmental impact of power to hydrogen/power to methane facilities on hybrid power-natural gas energy systems. *Int. J. Hydrogen Energy* **2020**, *45*, 20200–20209. [CrossRef]
148. Lovato, G.; Alvarado-Morales, M.; Kovalovszki, A.; Peprah, M.; Kougias, P.G.; Rodrigues, J.; Angelidaki, I. In-situ biogas upgrading process: Modeling and simulations aspects. *Bioresour. Technol.* **2017**, *245*, 332–341. [CrossRef]
149. Luo, Y.; Shi, Y.; Li, W.; Cai, N. Synchronous enhancement of H_2O/CO_2 co-electrolysis and methanation for efficient one-step power-to-methane. *Energy Convers. Manag.* **2018**, *165*, 127–136. [CrossRef]
150. Meylan, F.D.; Piguet, F.-P.; Erkman, S. Power-to-gas through CO_2 methanation: Assessment of the carbon balance regarding EU directives. *J. Energy Storage* **2017**, *11*, 16–24. [CrossRef]
151. Michailos, S.; Walker, M.; Moody, A.; Poggio, D.; Pourkashanian, M. A techno-economic assessment of implementing power-to-gas systems based on biomethanation in an operating waste water treatment plant. *J. Environ. Chem. Eng.* **2020**, *9*, 104735. [CrossRef]
152. Momeni, M.; Soltani, M.; Hosseinpour, M.; Nathwani, J. A comprehensive analysis of a power-to-gas energy storage unit utilizing captured carbon dioxide as a raw material in a large-scale power plant. *Energy Convers. Manag.* **2020**, *227*, 113613. [CrossRef]
153. Monzer, D.; Rivera-Tinoco, R.; Bouallou, C. Investigation of the Techno-Economical Feasibility of the Power-to-Methane Process Based on Molten Carbonate Electrolyzer. *Front. Energy Res.* **2021**, *9*, 195. [CrossRef]
154. Morgenthaler, S.; Ball, C.; Koj, J.C.; Kuckshinrichs, W.; Witthaut, D. Site-dependent levelized cost assessment for fully renewable Power-to-Methane systems. *Energy Convers. Manag.* **2020**, *223*, 113150. [CrossRef]
155. Mulat, D.G.; Mosbæk, F.; Ward, A.J.; Polag, D.; Greule, M.; Keppler, F.; Nielsen, J.L.; Feilberg, A. Exogenous addition of H_2 for an in situ biogas upgrading through biological reduction of carbon dioxide into methane. *Waste Manag.* **2017**, *68*, 146–156. [CrossRef] [PubMed]
156. Ortiz, S.; Rengifo, C.; Cobo, M.; Figueredo, M. Packed-bed and Microchannel Reactors for the Reactive Capture of CO_2 within Power-to-Methane (P2M) Context: A Comparison. *Comput. Aided Chem. Eng.* **2020**, *48*, 409–414. [CrossRef]
157. Patterson, T.; Savvas, S.; Chong, A.; Law, I.; Dinsdale, R.; Esteves, S. Integration of Power to Methane in a waste water treatment plant—A feasibility study. *Bioresour. Technol.* **2017**, *245*, 1049–1057. [CrossRef] [PubMed]
158. Pieta, I.; Lewalska-Graczyk, A.; Kowalik, P.; Antoniak-Jurak, K.; Krysa, M.; Sroka-Bartnicka, A.; Gajek, A.; Lisowski, W.; Mrdenovic, D.; Pieta, P.; et al. CO_2 Hydrogenation to Methane over Ni-Catalysts: The Effect of Support and Vanadia Promoting. *Catalysts* **2021**, *11*, 433. [CrossRef]
159. Sánchez, A.; Martín, M.; Zhang, Q. Optimal Design of Sustainable Power-to-Fuels Supply Chains for Seasonal Energy Storage. *Energy* **2021**, *234*, 121300. [CrossRef]
160. Savvas, S.; Donnelly, J.; Patterson, T.; Chong, Z.S.; Esteves, S.R. Methanogenic capacity and robustness of hydrogenotrophic cultures based on closed nutrient recycling via microbial catabolism: Impact of temperature and microbial attachment. *Bioresour. Technol.* **2018**, *257*, 164–171. [CrossRef]
161. Schlautmann, R.; Böhm, H.; Zauner, A.; Mörs, F.; Tichler, R.; Graf, F.; Kolb, T. Renewable Power-to-Gas: A Technical and Economic Evaluation of Three Demo Sites within the STORE&GO Project. *Chemie Ingenieur Technik* **2021**, *93*, 568–579. [CrossRef]
162. Sinóros-Szabó, B.; Zavarkó, M.; Popp, F.; Grima, P.; Csedő, Z. Biomethane production monitoring and data analysis based on the practical operation experiences of an innovative power-to-gas benchscale prototype. *Acta Agrar. Debr.* **2018**, 399–410. [CrossRef]
163. Stangeland, K.; Kalai, D.; Li, H.; Yu, Z. CO_2 Methanation: The Effect of Catalysts and Reaction Conditions. *Energy Procedia* **2017**, *105*, 2022–2027. [CrossRef]
164. Straka, P. A comprehensive study of Power-to-Gas technology: Technical implementations overview, economic assessments, methanation plant as auxiliary operation of lignite-fired power station. *J. Clean. Prod.* **2021**, *311*, 127642. [CrossRef]
165. Vo, T.T.; Rajendran, K.; Murphy, J.D. Can power to methane systems be sustainable and can they improve the carbon intensity of renewable methane when used to upgrade biogas produced from grass and slurry? *Appl. Energy* **2018**, *228*, 1046–1056. [CrossRef]
166. Wang, L.; Pérez-Fortes, M.; Madi, H.; Diethelm, S.; Van Herle, J.; Maréchal, F. Optimal design of solid-oxide electrolyzer based power-to-methane systems: A comprehensive comparison between steam electrolysis and co-electrolysis. *Appl. Energy* **2018**, *211*, 1060–1079. [CrossRef]
167. Wang, L.; Zhang, Y.; Pérez-Fortes, M.; Aubin, P.; Lin, T.-E.; Yang, Y.; Maréchal, F.; Van Herle, J. Reversible solid-oxide cell stack based power-to-x-to-power systems: Comparison of thermodynamic performance. *Appl. Energy* **2020**, *275*, 115330. [CrossRef]
168. Welch, A.J.; Digdaya, I.A.; Kent, R.; Ghougassian, P.; Atwater, H.A.; Xiang, C. Comparative Technoeconomic Analysis of Renewable Generation of Methane Using Sunlight, Water, and Carbon Dioxide. *ACS Energy Lett.* **2021**, 1540–1549. [CrossRef]
169. Xie, J.; Peng, Y.; Wang, X.; Yang, Y. Optimization on Combined Cooling, Heat and Power Microgrid System with Power-to-gas Devices. In Proceedings of the 4th Conference on Energy Internet and Energy System Integration (EI2) 2020, Wuhan, China, 30 October–1 November 2020; IEEE: Piscataway, NJ, USA, 2020; pp. 1862–1866. [CrossRef]
170. Zoss, T.; Dace, E.; Blumberga, D. Modeling a power-to-renewable methane system for an assessment of power grid balancing options in the Baltic States' region. *Appl. Energy* **2017**, *170*, 278–285. [CrossRef]

Article

Power-to-Gas and Power-to-X—The History and Results of Developing a New Storage Concept

Michael Sterner [1,2,*] **and Michael Specht** [3,4]

1. OTH Regensburg, 93053 Regensburg, Germany
2. Formerly Fraunhofer IEE (ISET, IWES), 34119 Kassel, Germany
3. Specht-eFuels, 71111 Waldenbuch, Germany; michael.specht@web.de
4. Formerly ZSW Stuttgart, 70563 Stuttgart, Germany
* Correspondence: michael.sterner@oth-regensburg.de

Abstract: Germany's energy transition, known as 'Energiewende', was always very progressive. However, it came technically to a halt at the question of large-scale, seasonal energy storage for wind and solar, which was not available. At the end of the 2000s, we combined our knowledge of both electrical and process engineering, imitated nature by copying photosynthesis and developed Power-to-Gas by combining water electrolysis with CO_2-methanation to convert water and CO_2 together with wind and solar power to synthetic natural gas. Storing green energy by coupling the electricity with the gas sector using its vast TWh-scale storage facility was the solution for the biggest energy problem of our time. This was the first concept that created the term 'sector coupling' or 'sectoral integration'. We first implemented demo sites, presented our work in research, industry and ministries, and applied it in many macroeconomic studies. It was an initial idea that inspired others to rethink electricity as well as eFuels as an energy source and energy carrier. We developed the concept further to include Power-to-Liquid, Power-to-Chemicals and other ways to 'convert' electricity into molecules and climate-neutral feedstocks, and named it 'Power-to-X' at the beginning of the 2010s.

Keywords: Power-to-Gas; Power-to-X; Power-to-Hydrogen; Power-to-Methane; hydrogen; methanation; sector coupling; sectoral integration; energy transition; eFuels; electric fuels; 100% renewable energy scenarios

1. Introduction

The energy transition is at the core of climate mitigation. Two-thirds of global greenhouse gas emissions result from the combustion of coal, oil and natural gas [1]. To move away from these stored fossil hydrocarbons, the expansion of renewable energy and energy efficiency are two fundamental steps. Among renewables, wind and solar energy show the greatest potential and lowest costs and have the lowest land consumption [2].

The core problem of wind and solar, however, is their intermittency. Flexibility options can solve this problem [3]:

1. Electricity networks can do spatial balancing but not temporal balancing;
2. Demand-side management can lower the storage demand;
3. Flexible power generation can react on wind and solar intermittency but requires stored energy carried in the form of hydrogen or hydrocarbons;
4. Storage is the most inefficient but only option to avoid blackouts and convert cheap wind and solar resources into storable energy carriers, fuels, feedstock and materials.

Storage technologies include short- and long-term storage technologies. Short-term storage technologies are characterized by high efficiencies, high cycling numbers and short discharge durations of a maximum 24 h. Examples are pumped hydro and batteries. Their weakness is high capacity costs and low energy density compared to hydrocarbons.

Additionally, batteries show much higher self-discharge rates than chemical storage sites such as gas caverns. Therefore, they do not solve the seasonal storage problem. Long-term storage facilities such as gas caverns show almost no self-discharge, low capacity costs and high energy density.

The main question in developing Power-to-Gas was how to access and use these vast storage capacities in the gas infrastructure for wind and solar. One option discussed was hydrogen in hydrogen caverns and fuel cells. However, these hydrogen technology components were either not available in the required scale and TRL or too expensive. Therefore, we copied photosynthesis, which does split water, whereby oxygen is released into the air. Nature, however, does not stop with hydrogen but combines it with CO_2 via direct air capture and thus generates CHO compounds in the form of biomass. This biomass is converted—after millions of years at high temperatures and pressures—into fossil energy carriers, which we use as the main storage to fuel almost everything, including the backup of wind and solar.

We simply copied these two core processes of photosynthesis technically by combining water electrolysis and CO_2-methanation, and 'Power-to-Gas' was born. What looks rather simple in retrospect was very challenging in the making. This development is given in this work to reflect and initiate similar innovations.

2. Method of Developing a New Storage Concept

The question of energy storage became increasingly urgent in Germany at the end of the 2000s, as renewable energy—especially wind and solar—experienced a broad introduction to the market via a proper regulatory framework. Since the beginning of energy balancing in energy economics, only a simple annual balance sheet has been sufficient due to the storable fossil energy sources used. For wind and solar, a simple annual value was also used. However, we conducted the first dynamic simulations of the electricity system with a high share of renewables on an hourly basis. This highlighted a great need for storage and balancing for the first time. It was clear that the identified demand could not be covered by existing storage methods in Germany such as batteries, compressed air storage or pumped hydro. Only pure hydrogen caverns were considered as an option as a solution [4,5]. Bioenergy was considered the only technical solution for balancing a 100% renewable electricity supply, as a hydrogen infrastructure was missing, and hydropower power was already exhausted to its potential limit in Germany.

In the 1990s, our colleagues at the Institute for Solar Energy Supply Technology (ISET) conducted the '250 MW wind turbine measurement program', from which they developed an hourly database of wind feed-in values all over Germany. This was the basis for developing wind power forecasts in the early 2000s, which were and still are essential for the grid integration of wind energy. Later, the first virtual power plant in the form of a wind farm cluster was created [6,7]. This all resulted in the 'Kombikraftwerk' project, which was able to demonstrate that a 100% renewable power supply is possible at any time on a scale of 1:10,000 in Germany. For this purpose, exemplary solar, wind and biogas power plants were combined and jointly controlled in real time to cover the virtual, downscaled power demand. The only facility that was simulated was the storage plant, represented by Germany's largest pumped hydro plant. This refuted assumptions that a 100% renewable power system is technically not possible and would cause blackouts and instability [4,8].

Analyses by Mackensen showed that for a fully renewable energy supply, mainly wind power and photovoltaics, would come into play and that these would require massive compensatory measures in the form of biomass or large storage capacities, which could be realized neither by adding pumped storage in Germany nor by the available areas for biogas [8,9]. The core challenge was the realistic upscaling of existing storage and biogas plants by 10,000 times. One solution proposed by ISET was the coupling of the electricity and gas sectors to store hydrogen from electrolysis with wind and solar electricity in the natural gas grid and flexibly convert it back into electricity via gas-fired power plants and CHP units [10].

In the same period, i.e., the end of the 1990s, we (Bandi, Weimer, Specht) and colleagues at the Center for Solar Energy and Hydrogen Research in Stuttgart (ZSW) developed a technological way to generate methanol from solar water electrolysis and atmospheric CO_2. The implemented pilot plant extracted CO_2 from the air via CO_2 absorption in a caustic air scrubber and electrodialysis for the regeneration process and, together with hydrogen from solar-powered electrolysis, converted it to methanol in a fixed-bed reactor filled with catalyst. This successfully demonstrated the technical feasibility of CO_2 recycling for methanol production [11–14]. ZSW focused on biomass gasification in the 2000s by developing a proprietary process, the Absorption-Enhanced Reforming Process (AER). In this way, we obtained a hydrogen-rich product gas from biomass via two coupled fluidized bed reactors and enhanced our knowledge of hydrogen-based fuels [15].

The scientific debate soon revealed that the potential of biomass remains limited. In 2008, the short-term rise in food prices caused the 'food or fuel' debate, and the choice between the use of biomass for energy or for food and fodder was a vivid debate that continues to shape public perception of biofuels today [16,17]. Bioenergy is very good for balancing intermittent wind and solar power but does not have the necessary sustainable potential [10]. We (Schmid, Sterner) at ISET concluded that the limited biomass resources are best integrated into our energy systems via gasification, fermentation and methanation as converted methane gas, which is fed into existing natural gas infrastructure [10,18]. There, the necessary transport and storage capacities are available, and gas is accessible to all energy sectors via boilers, CHPs, power plants and vehicles. At the same time, this use of bioenergy in the natural gas grid offers the possibility of capturing CO_2 and establishing a carbon sink. The sustainability of Bio-CCS (carbon capture and storage) is, however, discussed controversially, as underground sites are needed instead for renewable gases like hydrogen and SNG. Using this pathway, also fossil coal could be converted to fossil SNG, leaving the same problems with CCS.

The integrated energy system we designed in 2008 ultimately consisted of a coupled electricity and gas system with a possible CO_2 sink. The well-established conversion 'Gas-to-Power' was done by CHPs or gas-fired power plants. The new conversion from 'Power-to-Gas' was done by and electrolyzer to generate 'green hydrogen' for fuel cells and CHPs. The reformer was an optional way to convert natural gas into 'blue hydrogen' and store the remaining CO_2 underground (Figure 1).

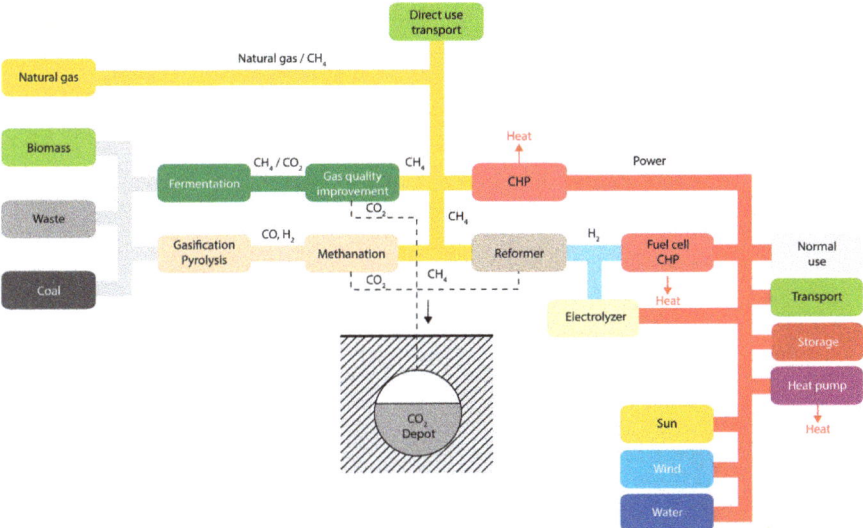

Figure 1. First approach to couple power and gas infrastructure via electrolysis; first published in our bioenergy flagship report at the German Advisory Council on Global Change (WGBU) [10].

This system was presented by Schmid and Sterner at the 16th European Biomass Conference [18,19]. Additionally present was Specht, who presented ZSW's work on biomass conversion to hydrogen [15]. We met after our presentations and discussed Specht's idea of using the Sabatier process via the methanation of CO_2 in Figure 1 instead of the reformer for the better integrability of hydrogen in the natural gas grid.

Hereby, the preliminary work of both of our institutes, ISET and ZSW, converged, and the idea was elaborated into the Power-to-Gas concept (Figure 2). This resulted in a patent application in 2009 [20], a first PhD thesis on PtG [21] and the development of the first PtG plant for CO_2 methanation in Germany on behalf of Gregor Waldstein from his SolarFuel GmbH [22].

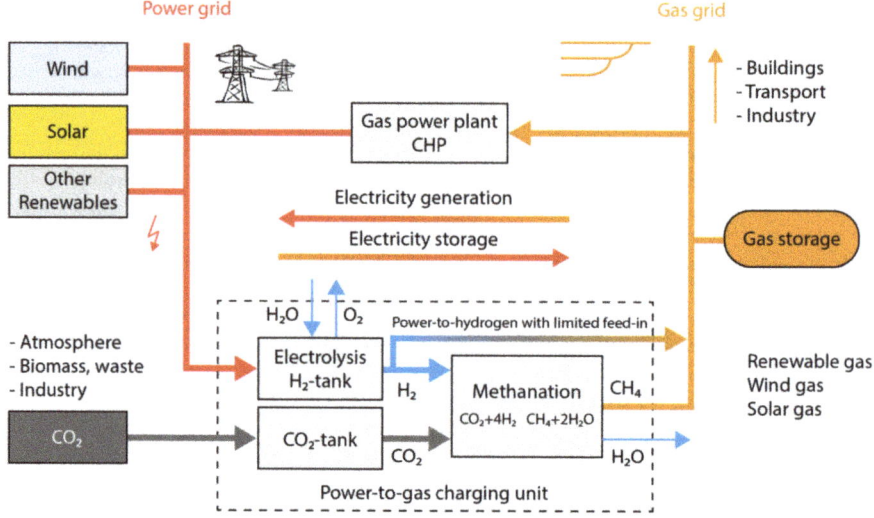

Figure 2. Basic concept of Power-to-Gas from 2008, first published in 2009 [20].

Subsequently, various pilot projects were jointly developed. In Kassel, we conducted studies for the energy and automotive industry (Uniper, Greenpeace Energy, Audi), and in Stuttgart, the hardware was further developed. By publishing and disseminating this new concept and implementing pilot projects, these two institutes enabled the idea of Power-to-Gas to achieve a breakthrough [23]. The largest PtG project realized to date with an electrical input power of 6 MW was implemented by Audi AG to run 1500 CO_2-neutral vehicles on PtG (see Section 3.9).

3. Results of the Development

3.1. The Original Power-to-Gas Concept

Power-to-Gas (PtG, P2G) describes the conversion of renewable electricity to renewable gas. Two core processes are combined: water electrolysis and CO_2 methanation. Renewable electricity drives the water splitting in the electrolysis. The resulting hydrogen is converted with CO_2 into methane in the methanation process. Methane is the main constituent of natural gas and thus the generated renewable gas is a substitute natural gas (SNG), that can be fed and stored 1:1 in the natural gas grid.

Power-to-Gas refers to a simple technical replica of the natural photosynthesis process in plants. These plants have developed the process over millions of years to be able to store solar energy over long periods of time. Regardless of its comparatively low efficiency of approximately one percent for solar irradiation to bioenergy, photosynthesis has proven itself in evolution. CO_2 and water are converted to compounds containing hydrogen, carbon and oxygen (e.g., $C_6H_{12}O_6$) in two steps using solar energy, and oxygen is released into the air.

The first step of photosynthesis, the splitting of water, is mimicked by PtG via electrolysis, with alkaline and membrane electrolysis available today and high-temperature electrolysis being developed. In the second step, H_2 reacts with CO_2, which is ideally taken from the atmosphere. Two processing options are available today for CO_2-methanation: chemical and biological methanation. Biogas or wastewater treatment plants, direct air capture (DAC), geothermal sources, fossil power plants or industrial processes serve as CO_2 sources.

The renewable methane gas can be fed into the gas network or stored in connected gas storage sites. From there it can be used for the transport or heat sector or converted back into electricity via gas turbines, CHP, or other devices such as fuel cells (Figure 2). The carbon cycle is closed: the CO_2 released to the atmosphere during combustion is the same that was previously extracted from it. A discussion of these CO_2 sources and their climate impact is given in Section 4.

3.2. Power-to-Gas: Coupling Electricity and Gas Sectors for Energy Storage

We were able to solve the chicken-and-egg problem of hydrogen by making the gas infrastructure, including transport, storage and applications, accessible for hydrogen via CO_2-methanation in Power-to-Gas. Hydrogen has only one-third the energy density of natural gas, dilutes the energy density of the gas accordingly and requires higher compression lines for transport and storage. Methane is easier to compress, store and transport. In addition, hydrogen injection was and is limited to low, single-digit percentages by limitations in gas turbines, gas tanks in vehicles, pore storage and material constraints. Through CO_2-methanation, we tapped the entire gas infrastructure for renewable electricity without limitations.

The development of this concept also marked the origin of the term 'sector coupling' or 'sectoral integration', which refers to energy storage via coupling of electricity and gas sectors. This results in the following opportunities:

- Fluctuating renewable energy can be stored seasonally.
- The existing gas infrastructure can store large TWh-amounts of renewable energy and transport it decoupled in time from the electricity grid all over Europe. This is an opportunity that the electricity infrastructure does not have on this scale.
- By converting the gas back into electricity, Power-to-Gas acts as electricity storage.
- CO_2 from biogas plants or other unavoidable sources finds a useful use as a carrier material for hydrogen.
- Process waste heat from all units can be used internally or via heat networks.
- Renewable gas can be generated anywhere and transported, distributed, stored and used without political or geographical constraints.
- Renewable gas can be used for heat supply to couple the electricity and heat sectors.
- Synthetic fuel can be used in mobility to couple the electricity and transport sectors.

3.3. The Chemistry behind Power-to-Gas

Electrolytic water splitting has been known for over 200 years and is therefore not a fundamentally new technology. Nevertheless, it is gaining importance in the context of PtG and PtX and is becoming the core component of chemical energy storage. Water is decomposed into hydrogen and oxygen using electrical energy (see Equation (1)).

$$2\ H_2O_{(l)} \rightarrow 2\ H_{2\ (g)} + O_{2\ (g)} \cdot \Delta H_R = 286\ \text{kJ/mol} \quad (1)$$

Two reversible equilibrium reactions occur in CO_2 methanation: the reverse water gas shift reaction and the CO methanation [24,25]. The first chemical reaction is responsible for separating the very weakly reactive CO_2 and occurs before the methanation reaction itself (see Equation (3)).

$$H_2 + CO_2 \rightarrow CO + H_2O_{(g)} \quad \Delta H_R = 41 \text{ kJ/mol} \qquad (2)$$

The second chemical reaction is the main reaction in which CO is hydrogenated (see Equation (3)). The CO methanation is as follows:

$$3H_2 + CO \rightarrow CH_4 + H_2O_{(g)} \quad \Delta H_R = -206 \text{ kJ/mol} \qquad (3)$$

Thus, the overall reaction for CO_2 methanation is Equation (4):

$$4H_2 + CO_2 \rightarrow CH_4 + 2H_2O_{(g)} \quad \Delta H_R = -165 \text{ kJ/mol} \qquad (4)$$

In reverse, this Sabatier reaction is known as steam reforming. This is the standard process for obtaining 'grey hydrogen' from fossil gas. The hydration of CO and CO_2 is strongly exothermic and volume-reducing, so the principle of Le Chatelier favors the methanation reactions at low temperatures and high pressures. A thermal management that reliably dissipates the released energy is therefore essential to keep the methanation reaction within a favorable temperature range and to shift the reaction equilibrium toward methane. This waste heat can be also used efficiently in other parts of the process, e.g., for the removal of CO_2 from biogas or air.

CO_2 methanation can be implemented both chemically and biologically in terms of process technology. We compared both in a standardization approach [26]. The biological route uses much lower temperatures and pressures, is more robust and less sensitive to gas impurities of the reactants than the chemical route, but is therefore mainly suitable for decentralized processes, especially in connection with biogas. Chemical methanation has long been proven, requires less space, has higher space-time yields, and is also available at large MW scales and offers a higher waste heat temperature level.

3.4. Novelty of CO_2 Methanization and Utilization in Energy Systems

Despite its discovery in France as early as 1902 from Paul Sabatier [27], CO_2-methanation was not explored for energy technology until much later, since, analogously to CO methanation, there was no need for it due to the cheap fossil resources available. In the 1970s and 1980s, storing solar energy chemically via CO_2 was discussed in terms of the 'SolChem Concept' in the USA [28,29]. On a laboratory scale, the first work and plants for CO_2 chemical methanation occurred in Japan in the 1990s, as Japan was a resource-poor and densely populated country conducting research on LNG power plants [30,31]. In Germany, CO_2 methanation was discussed in the context of fuel cells in the early 2000s [25]. Therefore, by our rethinking of energy systems, the usage of renewable electricity, water plus CO_2 as feedstock for renewable fuels, and using the Sabatier process for that particular purpose became popular after we introduced P2G [11,21].

3.5. Combining Electrical, Process and System Engineering Gives Interdisciplinary Solutions

We developed the processing technology that can be used to produce liquid or gaseous hydrocarbons from hydrogen and CO_2. To imitate nature, we used compounds of carbon and hydrogen as a storage medium to constantly cover our energy demand with natural sources. The technology was a new phenomenon: we used wind, solar, water and air to generate renewable fuels with the same quality as fossil fuels.

The power supply of an industrialized country such as Germany can therefore be met entirely with renewable energy, despite the natural fluctuations of photovoltaic, wind and hydroelectric power plants. With Power-to-Gas, the storage problem has been technically solved, and we can replace fossil fuels with renewables.

3.6. Terminology: Wind-to-Gas, e-Gas, Power-to-Hydrogen and Power-to-Methane

First, we named the concept 'wind-to-gas', then 'windgas' and 'solargas', to indicate the origin of this renewable gas and distinguish it from fossil natural gas. Similar was the case for 'renewable power methane' (RPM) and 'Renewable Power-to-Methane' (RPtM) or 'electric gas' (e-gas). The most fitting term would have been 'real natural gas' (RNG), but instead, inspired by Biomass-to-Liquid (BtL) and Gas-to-Liquid (GtL), we choose 'Power-to-Gas' (PtG, P2G).

The term Power-to-Gas became so popular that it was also used for hydrogen starting around 2012. Thus, the terms Power-to-Hydrogen (PtH, P2H) and Power-to-Methane (PtM, P2M) emerged to distinguish both processes. Power-to-Hydrogen describes the classical water electrolysis and the sector coupling via pure hydrogen. Power-to-Methane describes the classical route of Power-to-Gas.

3.7. Efficiencies

Besides the need for a CO_2 source, Power-to-Gas concepts differ in efficienciy. A complete Power-to-Gas storage system consists of a transformer, an electrolyzer, an optional methanation unit, compression and gas storage and a discharge technology, which varies according to the sectoral application of the stored gas.

The indicated efficiencies are mean values, from which different overall efficiencies result (Table 1). Regarding Power-to-Hydrogen storage systems, the total efficiency is about 5–12% higher than in the variants with methanation due to the lack of an intermediate methanation step.

Table 1. Efficiency chains (LHV) for different P2G applications, based on standard industry electrolysis technologies (alkaline and membrane (PEM)) and chemical methanation without energy demand for CO_2 provision and balance of plant. * compression to 80 bars [3,23].

Pathway	Overall Efficiency	Boundary Condition
Power-to-Hydrogen	54–72%	Compression to 200 bars (gas storage)
	57–73%	Compression to 80 bars (gas grid)
	64–77%	Without compression
Power-to-Methane	49–64%	Compression to 200 bars (gas storage)
	50–64%	Compression to 80 bars (gas grid)
	51–65%	Without compression
P2H-to-Power	34–44%	Power generation via fuel cell (60%) * or
P2M-to-Power	30–38%	combined-cycle power plant (60%) *
P2M-to-Heat and Power	43–54%	CHP (45% heat and 40% electricity) *
P2M-to-Heat	53–82%	Condensing boiler (105%) *
P2H-to-EnginePower	34–44%	Conversion in fuel cell (60%) *
P2M-to-EnginePower	18–22%	Combustion in gas engine (35%) *

3.8. Costs

The production cost of renewable gas and all other green C-based fuels is predominantly driven by investment costs, operating costs of the PtG/PtX-plants and the operation hours per year. Our findings are as follows. The investment costs fell very sharply within a decade and are still falling, as our published database shows (see Table 2). This is because the process plants have so far been manufactured mostly by hand. Investment costs can therefore fall sharply by plant automation and industrial production, similar to photovoltaics [2]. The operating costs are mainly related to renewable electricity, which is becoming cheaper and cheaper, and to taxes, levies and surcharges, which vary greatly and have been the biggest obstacle to the market introduction of PtG over the past decade.

Table 2. Average investment costs for the core components of PtG: alkaline and membrane electrolysis (alkaline, membrane (PEM)) and methanation (chemical, biological) for the MW class. The kW unit refers to the electrical power input of the electrolysis, not the gas flow rate [3].

Year	Alkaline Electrolysis in EUR/kW	Membrane Electrolysis in EUR/kW	Chemical Methanation in EUR/kW	Biological Methanation in EUR/kW
2010	1150	1650	1040	1600
2015	980	1350	870	1300
2020	850	1130	740	1050
2025	720	950	620	860
2030	620	780	520	690
2040	460	530	370	460
2050	330	350	260	300

What all chemical plants have in common is that profitability with the high investment costs requires operation at high utilization rates. Three core factors favor the economic operation of PtG/PtX-plants:

1. low-cost, renewable electricity;
2. high capacity factors/utilization rates;
3. favorable regulatory frameworks due to no or low charges, taxes and levies.

3.9. Advantages and Opportunities

Power-to-Gas enables a bidirectional coupling of electricity and gas grids. This is the greatest opportunity: to use the convergence of these systems for a sustainable energy supply with electricity, heat and fuel on the basis of wind and solar electricity using the existing networks and infrastructures for distribution and storage.

In addition to the already existing and huge storage natural gas grid, the great advantage of PtG is the versatile use of methane: unlike pure electricity storage plants such as pumped hydro or batteries, the injected gas does not necessarily have to go back into the power grid at the end but can be used in multiple ways and in multiple places. The stored energy is not fixed locally, as it is the case with pumped hydro or batteries. Seasonal storage can be implemented: the energy collected during sunny and windy seasons can be used in the winter or next spring for completely different purposes and at any location in the natural gas network—for heating, for mobility or even for reverse power generation in one of the many combined heat and power plants. This is not possible with battery or pumped storage: if they had to store the stored energy for a few weeks or even half a year, they would immediately become uneconomical, and some batteries discharge themselves within this time. Additionally, they can only return the power as electricity, and only at the same location.

This does not mean that these storage facilities are worse than Power-to-Gas. In fact, they are twice as efficient if used exclusively as power-to-power storage. However, they are far less flexible and essentially suitable for short-term day-night storage of electricity and for balancing short-term electricity peaks or deficits. Therefore, they do not compete with Power-to-Gas but are an important complement. These different fields of application are shown in Figure 3 based on storage capacities in Germany. Power-to-Gas thus plays a key role in the goal of leveraging synergies by coupling the electricity, heat and mobility sectors, and thus has a special position among storage technologies.

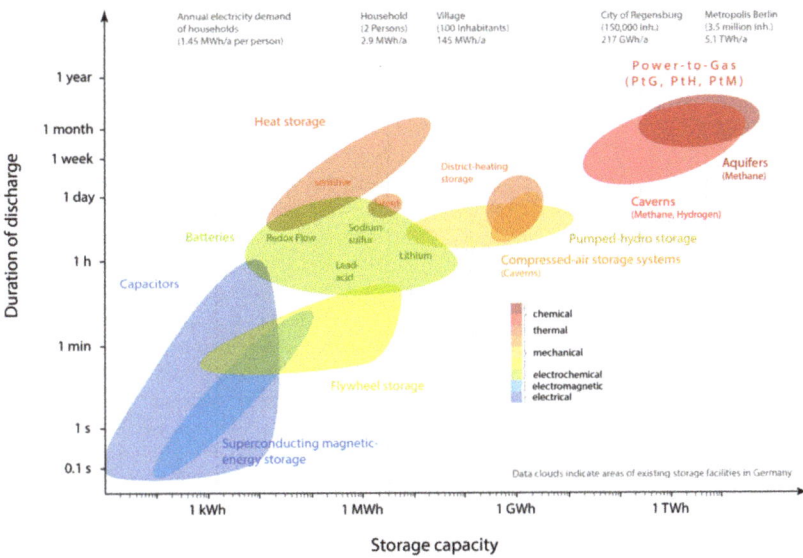

Figure 3. Storage capacities and duration of discharge of various storage technologies [3].

3.10. Disadvantages and Challenges

Power-to-Hydrogen is preferable for reasons of cost and efficiency as long as hydrogen can be stored and used locally, or the blending limit of the gas grid is not reached. If the fossil gas flow rate is low, the injection limit is quickly reached, and hydrogen buffer storage must be used to level the hydrogen injection.

In contrast to methane or natural gas, there are still no mass- and area-wide solutions for some applications. For example, although fuel cells have long been in research and development in heat and transport sectors, a market-ready technology on the required scale is not yet available.

Adapting natural gas infrastructure to higher hydrogen blending involves research and high costs. So far, 2% by volume is permitted in Germany, and 20% is considered technically possible. If all components of a pure hydrogen economy are affordable and available at a sufficient scale, methanation will become obsolete. However, if this is not yet the case, the existing gas infrastructure can be used for renewable gas.

3.11. Dissemination via Energy Economy Studies

In the energy industry, science and ministries, the concept became known through our work at Fraunhofer IEE (formerly ISET) based on simulations in major studies of the long-term scenario 'Lead Study' of the Federal Ministry for the Environment (BMU) [32], the German Advisory Council on the Environment (SRU) report [33], the 100% Renewable Electricity Target 2050 of the Federal Environmental Agency (UBA) [34] and the storage study of the Association of German Electrical Engineers (VDE) [35]. We also integrated the concept of P2G and sector coupling in the IPCC special report on renewable energy (SRREN), which made it familiar to the international scientific community [36].

We modeled the entire energy system with new findings in these studies: the existing natural gas grid in Germany is sufficient to buffer electricity surpluses with its large network and underground storage facilities.

For example, the UBA 100% renewable scenario showed a stable electricity supply with no blackouts, where 80% of the electricity demand is covered by wind, solar and hydropower. The remaining 20% of electricity demand were met with pumped hydro, batteries, and Power-to-Gas via gas storage and via CCGT power plants. The results were as follows:

- A full supply with renewable energy in all sectors is technically and ecologically feasible in 2050.
- The technical potential for onshore wind (390 TWh), offshore wind (260 TWh) and photovoltaics (250 TWh) is capable of meeting the energy demand for electricity, heat and individual mobility.
- Despite a very high installation of wind and solar with a total capacity of 225 GW and a peak load in Germany's power supply of about 80 GW, a reserve power (backup) of about 60 GW is needed; gas power plants based on PtG.
- The security of supply is ensured by PtG storage and gas turbines and CHPs.
- Despite the ideal expansion of the electricity grids and the use of large load management potentials via heat pumps including heat storage (44 TWh), air conditioning (28 TWh) and the controlled charging of electric vehicles (50 TWh), 85 TWh of 150 TWh electricity surpluses remain, which must be integrated via storage.
- The potential of the short-term storage technology pumped hydro is fully utilized with 0.055 TWh. It allows peak shaving but is far from sufficient to cover the storage needs in a 100% renewable power supply.
- On the other hand, only 15% of the technical potential of gas storage facilities is required for this task of long-term storage: 75 of 514 TWh. By curtailing 1% of the surplus energy, the PtG capacity of 44 GW can be designed to meet 64% of the maximum surplus capacity. The possible additional one percent of energy storage would involve a disproportionately high technical and financial storage effort.

The existing gas storage potential in Germany is about 220 TWh. With reconversion via CCGT power plants, about 120 TWh of electricity can be generated from the stored gas quantities in purely balance terms, which corresponds to 20% of the annual electricity consumption in Germany. This coud close all gaps in a renewable electricity supply.

In theory, if the 44 million vehicles that exist in Germany today were simultaneously connected to the grid as electric vehicles with a capacity of 20 kWh, half of which can realistically be used to compensate for deficits in the power system, 0.44 TWh of storage capacity would be available. If discharged at 60 GW, all vehicles would be able to stabilize the power grid for 7 h; all gas storage with the same discharge capacity 2000 h, or about 3 months. This comparison shows the storage potential of PtG (Figure 4).

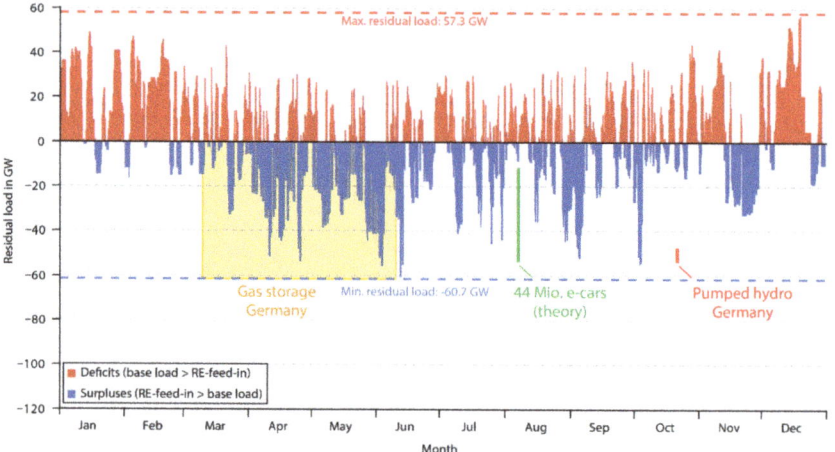

Figure 4. Deficits and surpluses in a 100% renewable electricity supply over the year 2050 in an example scenario for Germany and possible electricity storage options [3].

In addition to storage capacities, the gas grid also has a well-developed transport and distribution network. A large continental gas pipeline can transport energy in the form of gas with a capacity of about 70 GW, whereas standard electricity transmission lines have a typical electrical transmission capacity of 3.5 GW (two 380 kV three-phase systems).

Thus, PtG grid coupling not only enables the storage of large amounts of energy, but also a spatial shift in storage and the use of the renewable methane. This is one of the unique selling points of Power-to-Gas.

3.12. Dissemination via Demonstration Plants

The technical feasibility of PtG is demonstrated by numerous pilot plants in Europe and elsewhere. We published the current global status of PtG plants in 2019 [37].

Back in 2009, we decided to implement an initial exploitation of the idea of Power-to-Gas with our partner SolarFuel GmbH. On behalf of this new company, we built at ZSW the first PtG plant. This demonstrated the technical feasibility of the patented PtG technology and provided further insights into CO_2 methanation, which was largely unexplored in the energy sector. At Fraunhofer IWES, we explored the energy integration of PtG in accompanying research on optimized plant operation and concepts for high utilization rates of PtG plants at wind parks while simultaneously serving the electricity network operation via forecast balancing.

This alpha plant consists of two containers and uses the air as a CO_2 source (Figure 5). The first container contains a scrubber for CO_2 absorption in a scrubbing solution and an electrodialysis unit that expels the CO_2 from the scrubbing medium. The second container contains a 25 kW alkaline electrolyzer. A chemical fixed-bed reactor in pipes is used for methanation. A fuel maker is used to fuel gas cars with the produced synthetic natural gas (SNG) [38]. As the purpose of this first PtG pilot plant was to demonstrate the technology and concept, no process optimization was carried out, which means that the efficiency of the plant in converting electricity to gas is 40%.

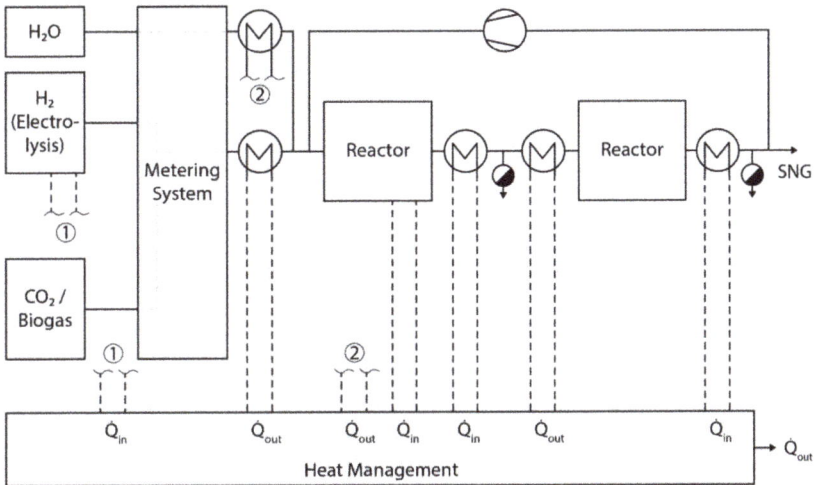

Figure 5. Block diagram of the first 25 kW PtG pilot plant at ZSW Stuttgart [38].

A 250 kW second plant was built for research purposes by the same consortium in Stuttgart with the support of the Federal Ministry of the Environment (Figure 6). Besides upscaling to the MW class, technical and economic research questions have been answered in the integration into the power grid and energy markets such as power balancing, load control, cost-optimized operation and sustainable CO_2 sources. A tube and a plate reactor were tested and compared as fixed-bed reactors. The plant supplied 50 m^3/h of hydrogen, which was converted into a gas output of 125 kW (LHV) via the two different routes of chemical methanation, corresponding to a product gas flow of 12.5 m^3/h [38].

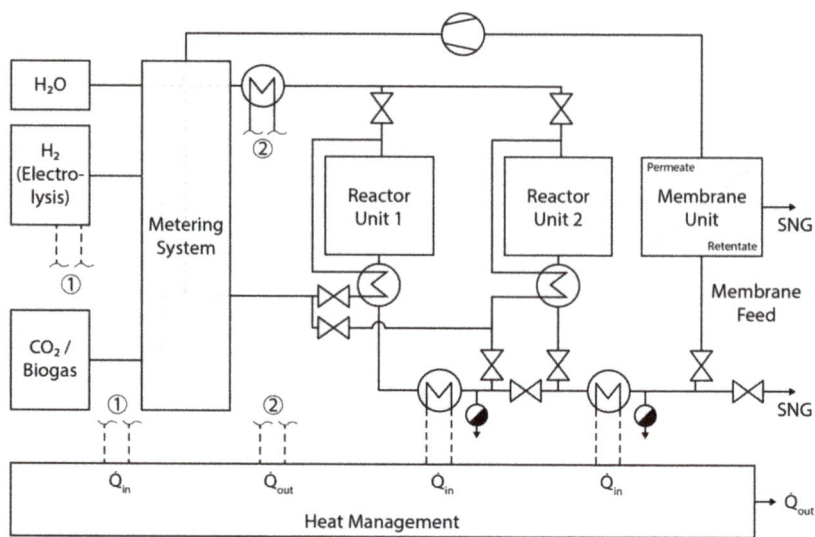

Figure 6. Block diagram of the second 250 kW PtG pilot plant at ZSW Stuttgart [38].

We applied the results to the Audi e-gas plant in Werlte in the world's first industrial PtG plant. Three alkaline electrolyzers with a total electrical input of 6 MW are used (Figure 7). The hydrogen produced is temporarily stored in a storage tank for up to one hour and compressed to 10 bars for the methanation stage [39]. The tube reactor operates as a fixed-bed reactor at temperatures of 200–350 °C and pressures of 5–10 bar with staged gas addition. Via a single-stage process, a nickel catalyst is used to achieve a methane quality > 90 vol.% CH_4. As a CO_2 source, biogas from a residual waste plant was used. This connection made it possible to use the waste heat from the electrolyzers and the methanation unit in the upgrading plant to sanitize the residual waste and to separate the CO_2 from the amine scrubbing liquid.

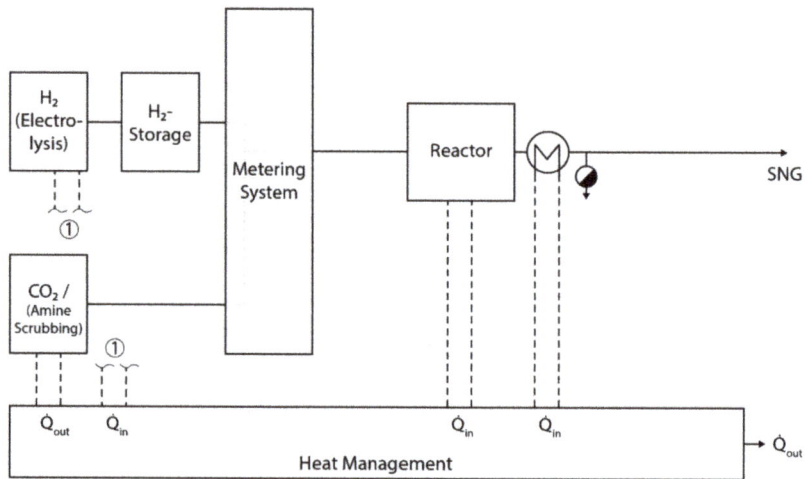

Figure 7. Block diagram of the resulting first commercial 6000 kW PtG plant at Audi AG [38].

Biogas can optionally be used directly as a CH_4/CO_2 blend in the methanation reactor without upgrading, as the gas cleaning worked well and the reactor catalyst tolerated this [40]. This can be a benefit for future plants, as one gas process unit can be saved. The plant was able to produce up to 1000 t of SNG per year, which can be used to refuel about 1500 Audi g-tron vehicles with an annual mileage of 15,000 km [38].

The project resulted from our first two pilot plants and was implemented by Audi AG together with MAN Diesel & Turbo, EWE AG, MT Biomethan GmbH and us at ZSW and IWES. The accompanying research was funded by the German Federal Ministry for the Environment, and the plant was inaugurated in Lower Saxony in June 2013 [38].

Many other projects in Germany and elsewhere followed. E-On erected a 2 MW P2H pilot plant 'WindGas' in Falkenhagen between Hamburg and Berlin. The main reasons for building the plant on the greenfield site were the high level of wind in Brandenburg and its proximity to the gas infrastructure. Back then it was the first plant to feed hydrogen directly into the gas transmission grid in Germany. The alkaline electrolyzers from Hydrogenics produced 360 m^3/h of hydrogen. The efficiency (LHV) of the total chain was 66%. It was operated by an E.ON subsidiary (now Uniper), and the gas was marketed to Switzerland via Swissgas AG. We co-initiated the project through feasibility studies in 2010 [41]. Later, a methanation unit was added in the EU project 'Store and go'. Other projects such as the hydrogen energy park in Mainz from Siemens AG, Linde AG and Stadtwerke Mainz AG with a newly developed 6 MW new proton exchange membrane electrolysis followed later in the same way.

Additionally, in 2013, Thüga AG, as a municipal utility consortium with 12 project partners, built the first P2H plant in Frankfurt for the gas distribution network. Here, ITM Power's 315 kW PEM electrolysis produced 60–70 m^3/h of 99.8 vol-% pure hydrogen from 4.9–5.2 kWh electricity per cubic meter. This hydrogen flow was mixed with a fossil gas flow of 3000 m^3/h in the gas network via a gas pressure control measurement and mixing system. The gas flow in the distribution network in the inner city of Frankfurt is constant over the year, allowing a low-dose injection of green hydrogen [39].

Enertrag's hybrid power plant in Prenzlau consists of a self-built 600 kW alkaline electrolysis plant, a hydrogen storage facility, a biogas plant, and a CHP unit. The 120 m^3/h hydrogens are directly marketed via trailers for transport and industry. The plant was inaugurated in 2011 and entered the test phase. It has been in operation since 2013, and since 2014, inspired by us, the hydrogen has been fed into the natural gas grid to be supplied to approximately 8000 customers of Greenpeace Energy eG. With the proWindgas gas tariff, they are the first energy supplier to promote PtG via direct marketing to customers to promote the energy transition towards a 100% renewable energy supply [41,42].

Our pilot projects and these other early-stage projects can be used as blueprints for future commercial plants.

3.13. From Power-to-Gas to Sector Coupling to Power-to-X–Definitions

Power-to-Gas is the cornerstone of sector coupling and sectoral integration:

- Electricity and gas sectors are coupled for seasonal energy storage;
- Electricity and heat sectors are linked via renewable gas;
- Electricity and transport sectors are coupled via synthetic fuel;
- Renewable gas serves as a link between the power and industrial sectors to make steel, chemicals and other sectors that are difficult to decarbonize climate-neutral (see Figure 8).

All at the same time, Power-to-Gas is used where pure electricity applications are not technically sufficient to make the respective sector completely climate-neutral.

We inspired so many other researchers, who came up with similar ideas by using renewable electricity as 'primary energy', that we decided in 2013/14 to summarize all terms to one: Power-to-X. We gave two new definitions, which became very popular over time:

Definition of Power fuels/eFuels: Power fuels or eFuels are chemical energy carriers based on electrical energy, produced via the electrolysis of water and an optional synthesis

(PtG, PtL) and used in mobility. Examples are hydrogen, SNG, methanol, ammonia and Fischer–Tropsch fuels like e-diesel or e-kerosene [41].

Definition of Power-to-X: Power-to-X describes the conversion and storage of electrical energy into an energy carrier (gas, fuel or raw material) or a product (basic material, feedstock). It is a collective term for Power-to-Gas, Power-to-Liquid, Power-to-Fuel, Power-to-Chemicals and Power-to-Product [3,43].

This completed the existing definitions of PtG, PtL, PtC, eFuel and sector coupling [41]:

Definition of Power-to-Gas: A Power-to-Gas (PtG) plant describes a facility for converting electrical energy into a gaseous energy carrier such as hydrogen or methane via water electrolysis and optional methanation and storing it. It is thus part of an energy storage system. Power-to-Gas describes on the one hand the plant for conversion and storage and on the other hand also the overall system, which consists of injection (electrolysis, methanation) storage (gas storage, gas grid) and withdrawal (gas power plants, CHP, gas mobility and gas heating).

Definition of Power-to-Liquid: A Power-to-Liquid (PtL) plant describes a plant for the conversion and storage of electrical energy into a liquid energy carrier such as kerosene, diesel or methanol via water electrolysis and syntheses. The energy carrier is used for energy.

Definition of Power-to-Chemicals: A Power-to-Chemicals (PtC) plant describes a plant for converting and storing electrical energy into a chemical product such as methanol via water electrolysis and syntheses. The product is used as a material.

Definition of eFuel: An eFuel is a liquid or gaseous fuel based on the conversion and storage of electrical energy via electrolysis and optional syntheses (Power-to-X). The energy carrier is used for energy.

Definition of sector coupling: Sector coupling describes the coupling of the electricity sector with the building, transport and industry sectors via, for example, CHP, heat pumps, heating rods, electromobility, Power-to-X using energy conversion, energy networks (gas, fuel, raw material, electricity) and energy storage.

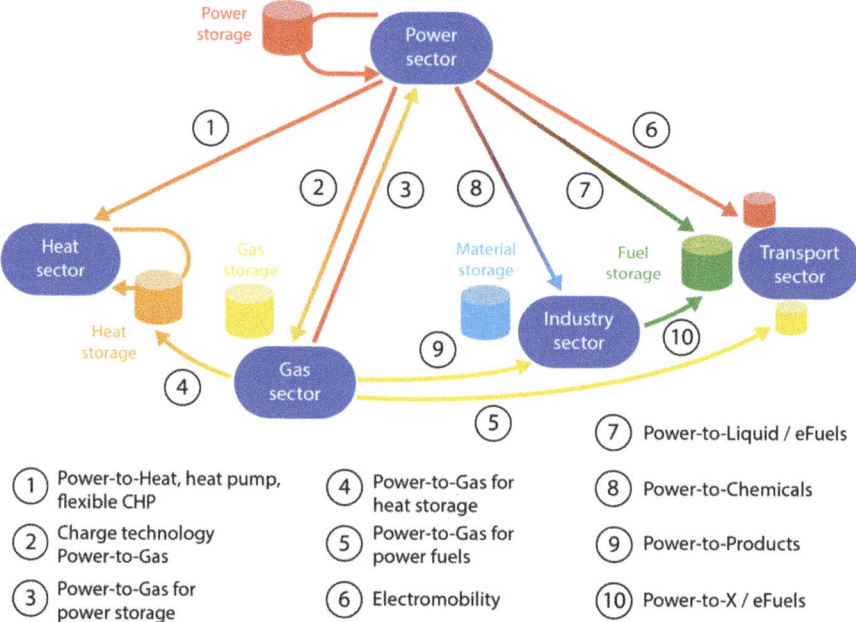

Figure 8. Sector coupling resulted from Power-to-Gas and Power-to-X. It links the sectors of electricity, heat, transport and industry via energy storage and energy converters and using renewable electricity as 'primary energy' for ecarbonization [3].

4. Discussion of Electricity- and C-Sources for the Generation of C-Based Fuels

4.1. Climate-Neutrality of CO_2 Based Fuels and Their Usage

The use of CO_2 in the process is climate-neutral overall unless fossil CO_2 is released specifically for this purpose. Fossil power generation does not become CO_2-neutral by the utilization of the produced CO_2, since the CO_2 is released back into the atmosphere after use (CCU—carbon capture and usage) in the PtG storage cycle. In addition, CO_2-rich gases such as those from biogas or biomass are more suitable as a source than flue gas from power plants, since the energy required to capture the CO_2 is lower [21].

4.2. From Fossil CO_2 to Direct Air Capture—C-Sources for C-Based Fuels

After biogas upgrading, pure CO_2 is available, which is cost-efficient and climate-neutral since it is normally unused and was removed from the atmosphere by plants within a very short period of time.

Another option for using biogenic CO_2 is the use of raw biogas, which is advantageous because the CO_2 does not have to be extracted separately. Hydrogen reacts with the CO_2 content of biogas directly in the methanation reactor. This enables a very broad and distributed use of the decentral renewable energy source via the central gas infrastructure, as opposed to rigid on-site electricity generation from biogas as a base load.

The obvious option is to use CO_2 directly from the air, which together with a CCS process would lead to a reduction in the CO_2 content of the air. Our first 25 kW P2G pilot plant uses separation from air via absorption in caustic liquids [22]. However, capture from the air is only possible with high technical and financial effort, which is why this option has not been explored fully. The advantage is site independence for CO_2 recovery, which is often a limiting or site-determining factor in the implementation of hydrogenation syntheses.

Large quantities of CO_2 are produced in power plants and industrial facilities. While fossil power plants can be replaced by renewable energies, for example, cement industries continue to inevitably emit CO_2. The same applies to other processes that are independent of the energy transition. However, unlike biogenic or atmospheric CO_2, these CO_2 sources are not distributed decentrally, which is an advantage for large-scale production of synthetic fuel such as 'e-kerosene'. Combining near-shore offshore wind farms, which generate constant and cheap electricity, with large CO_2 emitters such as cement plants results in the synergetic approach of generating 'unavoidable' jet fuel from 'unavoidable' CO_2 emissions.

Coal power plants are not considered as a source of CO_2, since in the short term, CCU with PtX is not intended to 'green-wash' CO_2-intensive power generation and in the long run, fossil power plants will disappear during the energy transition and along with them the CO_2 sources. Furthermore, the additional effort of flue gas treatment increases the primary energy consumption of power plants by 20–44% [44]. This means that part of the CO_2 emissions that are captured are caused by the increased energy input just because CO_2 is to be captured. On closer examination, this is paradoxical and only has a net effect if the CO_2 is stored for thousands of years or no longer escapes into the atmosphere when firmly bound via material use.

CO_2 recycling is understood to be a closed CO_2 cycle. After the combustion of renewable methane, CO_2 is separated from the flue gases and made available again for methanation. If pure oxygen, which is a by-product of electrolysis, is used in the combustion of the methane and burned under the correct stoichiometric ratio, CO_2 capture is possible without much effort, since only CO_2 and water are produced in such combustion.

4.3. Sustainable, Renewable C-Based Fuels Require Renewable Electricity and Proper Frameworks

In addition to the CO_2 source, the use of renewable electricity is the basic prerequisite for a reduction of CO_2 by Power-to-Gas: the CO_2 intensity of the electricity source determines the CO_2 intensity of the PtX product or hydrogen derivate. Choosing fossil electricity as an energy source for PtX reverses its climate effect: converting lignite into electricity into gas, and thus a fossil chemical energy carrier into an electrical one, and in the second step again back into a chemical one, we do not get a CO_2-neutral energy carrier.

The PtG gas would emit about 3200 g CO_2 eq./kWh and thus more than eight times more as fossil electricity from fossil gas when converted back into electricity [3].

From the perspective of sustainability, the use of renewable electricity is a must, and biogenic or atmospheric CO_2 sources are favored. The challenge and future main research question remain how to modify the regulatory framework in that way that renewable gas becomes competitive against natural gas [45]. CO_2 pricing, quota systems and OPEX funding are initial approaches towards the broad market introduction of PtX.

5. Conclusions

We created and patented a new storage concept called Power-to-Gas at the end of the 2000s that imitates photosynthesis and generates renewable hydrocarbons from renewable electricity, water and CO_2. This inspired others and led to the concepts of synthetic fuels (eFuels, Power-to-Liquids and Power-to-Chemicals). We summarized all power-to-concepts in the term Power-to-X in 2013. We built the first demo plants at a kW-scale, which led to the first MW-scale plants at the beginning of the 2010s and inspired many other PtX-plants worldwide.

The main advantages of PtG are that we (i) solved the chicken-egg-problem of H_2, (ii) solved the seasonal storage problem of wind and solar by coupling electricity and gas sectors and thus using existing infrastructures with TWh-scale storage for renewables and (iii) created a way to generate eFuels from wind and solar for transport, industry and buildings.

The main disadvantages are the need for a CO_2 source and the lower efficiency compared to the direct use of electricity or hydrogen. However, this is very useful for all applications that cannot decarbonize or otherwise become climate-neutral: long-distance transport like ship, airplanes and heavy-duty, 100% renewable electricity supply and many industry applications such as high-temperature processes.

6. Patents

We patented our Power-to-Gas concept innovation in the EU and USA:

1. Specht, M.; Sterner, M.; Stürmer, B.; Frick, V.; Hahn, B. Renewable Power Methane–Stromspeicherung durch Kopplung von Strom- und Gasnetz–Wind/PV-to-SNG. DE 10 2009 018 126.1, 9 April 2009.
2. Specht, M.; Sterner, M.; Stürmer, B.; Frick, V.; Hahn, B. Energy System and Supply Method, EP 00 0002 3345 90B1, 9 April 2010.
3. Specht, M.; Sterner, M.; Stürmer, B.; Frick, V.; Hahn, B. Energy System and Supply Method, US 00 0009 0571 38B2, 9 April 2010.

Author Contributions: Conceptualization, formal analysis, investigation, resources, data curation, visualization, project administration, funding acquisition, M.S. (Michael Sterner) and M.S. (Michael Specht); methodology, software, writing—original draft preparation, M.S. (Michael Sterner); validation, writing—review and editing, supervision, M.S. (Michael Specht). All authors have read and agreed to the published version of the manuscript.

Funding: This research was partly funded by the German Federal Environmental Agency in the research project 'Modeling of 100 percent renewable power generation in 2050', FKZ 363 01 277. The assembly and operation of the 25-kW P2G ® plant were financed by SolarFuel GmbH, renamed ETOGAS GmbH in 2013. The assembly and operation of the 250 kW P2G ® plant were supported by the Federal Ministry for Economic Affairs and Energy based on a resolution of the German Parliament (funded project no.: 0325275). The construction of the 6000 kW e-gas plant is an investment of the automobile manufacturer Audi AG. The monitoring of the e-gas plant is supported by the Federal Ministry for Economic Affairs and Energy based on a resolution of the German Parliament (funded project no.: 0325428).

Acknowledgments: We thank Franz Bauer, OTH Regensburg, for his technical support in drawing the figures and illustrations for this publication.

Conflicts of Interest: The authors declare no conflict of interest.

References

1. IEA. *World Energy Outlook*; OECD: Paris, France, 2020.
2. Kost, C.; Shammugam, S.; Fluri, V.; Peper, D.; Memar, A.; Schlegl, T. Levelized Cost of Electricity-Renewable Energy Technologies. *Fraunhofer Inst. Sol. Energy Syst. ISE* **2021**, *144*, 2–5.
3. Sterner, M.; Stadler, I. *Handbook of Energy Storage: Demand, Technologies, Integration*; Springer: Berlin/Heidelberg, Germany, 2018; ISBN 978-3-662-55503-3.
4. Sauer, D. The demand for energy storage in regenerative energy systems. In Proceedings of the First International Renewable Energy Storage Conference (Eurosolar IRES I), Gelsenkirchen, Germany, 30 October 2006.
5. VDE. *Energiespeicher in Stromversorgungssystemen mit Hohem Anteil Erneuerbarer Energieträger*; VDE (Verband der Elektrotechnik, Elektronik und Informationstechnik): Frankfurt, Germany, 2009.
6. Enßlin, C.; Füller, G.; Hahn, B.; Hoppe-Kilpper, M.; Rohrig, K. The Scientific Measurement and Evaluation Programme in the German 250 MW Wind Programme. In Proceedings of the European Wind Energy Conference, Kassel, Germany, 8–12 March 1993.
7. Ernst, B. Entwicklung eines Windleistungsprognosemodells zur Verbesserung der Kraftwerkseinsatzplanung. Ph.D. Thesis, University of Kassel, Kassel, Germany, 2003.
8. Mackensen, R.; Rohrig, K.; Emanuel, H.; Saint-Drenan, Y.; Schlögl, F. Das regenerative Kombikraftwerk: Abschlussbericht, Kassel. 2008. Available online: https://www.researchgate.net/profile/Kurt-Rohrig/publication/46112051_Das_regenerative_Kombikraftwerk/links/0deec517c0f740311b000000/Das-regenerative-Kombikraftwerk.pdf (accessed on 2 August 2021).
9. Mackensen, R. Herausforderungen und Lösungen für eine regenerative Elektrizitätsversorgung Deutschlands. Ph.D. Thesis, University of Kassel, Kassel, Germany, 2011.
10. WBGU. *World in Transition—Future Bioenergy and Sustainable Land Use: Flagship Report 2008*; WBGU: Berlin, Germany, 2009.
11. Bandi, A.; Specht, M.; Weimer, T.; Schaber, K. CO_2 recycling for hydrogen storage and transportation: Electrochemical CO_2 removal and fixation. *Energy Convers. Manag.* **1995**, *36*, 899. [CrossRef]
12. Specht, M.; Staiss, F.; Bandi, A.; Weimer, T. Comparison of the renewable transportation fuels, liquid hydrogen and methanol, with gasoline—energetic and economic aspects. *Energy Convers. Manag.* **1998**, *23*, 387–396. [CrossRef]
13. Specht, M.; Bandi, A.; Elser, M.; Heberle, A.; Maier, C.U.; Schaber, K.; Weimer, T. CO2-Recycling zur Herstellung von Methanol: Endbericht, Stuttgart. 2000. Available online: https://www.sfv.de/pdf/Report_000700_ZSW_CO2_to_MeOH_LQ2.pdf (accessed on 2 August 2021).
14. Weimer, T.; Schaber, K.; Specht, M.; Bandi, A. Methanol from atmospheric carbon dioxide: A liquid zero emission fuel for the future. *Energy Convers. Manag.* **1996**, *37*, 1351–1356. [CrossRef]
15. Specht, M.; Zuberbühler, U.; Koppatz, S.; Pfeifer, C.; Marquard-Möllenstedt, T. Transfer of Absorption Enhanced Reforming Process (AER) from Pilot Scale to an 8 MW Gasification Plant in Guessing, Austria. In Proceedings of the 16th European Biomass Conference & Exhibition' of EUBIA, Feria Valencia, Spain, 2–6 June 2008; p. 20.
16. FAO. *The State of Food and Agriculture 2008: Biofuels—Prospects, Risks and Opportunities*; FAO: Rome, Italy, 2008; ISBN 9251059802.
17. Sachs, J.D. Surging food prices mean global instability: Misguided policies favor biofuels over grain for hungry people. *Sci. Am. Mag.* **2008**, *2*. Available online: https://www.scientificamerican.com/article/surging-food-prices/ (accessed on 2 August 2021).
18. Schmid, J.; Sterner, M. Bioenergy in future energy systems. In Proceedings of the 16th European Biomass Conference & Exhibition' of EUBIA, Feria Valencia, Spain, 2–6 June 2008; p. 26.
19. Sterner, M.; Schmid, J. Electromobility—An efficient alternative to conventional biofuels to put biomass on the road. In Proceedings of the 16th European Biomass Conference & Exhibition' of EUBIA, Feria Valencia, Spain, 2–6 June 2008; p. 16.
20. Specht, M.; Sterner, M.; Stuermer, B.; Frick, V.; Hahn, B. Energieversorgungssystem und Betriebsverfahren. Germany Patent 10 2009 018 126.1, 9 April 2009.
21. Sterner, M. Bioenergy and Renewable Power Methane in Integrated 100% Renewable Energy Systems: Limiting Global Warming by Transforming Energy Systems. Ph.D. Thesis, University of Kassel, Kassel, Germany, 2009.
22. Specht, M.; Sterner, M.; Brellochs, J.; Frick, V.; Stuermer, B.; Zuberbühler, U.; Waldstein, G. Speicherung von Bioenergie und erneuerbarem Strom im Erdgasnetz: Storage of Renewable Energy in the Natural Gas Grid. *Erdöl Erdgas Kohle* **2010**, *126*, 342–346.
23. Sterner, M.; Jentsch, M.; Holzhammer, U. *Energiewirtschaftliche und Ökologische Bewertung eines Windgas-Angebotes*; Gutachten: Hamburg, Kassel, 2011. [CrossRef]
24. Kopyscinski, J.; Schildhauer, T.J.; Biollaz, S.M. Production of synthetic natural gas (SNG) from coal and dry biomass—A technology review from 1950 to 2009. *Fuel* **2010**, *89*, 1763–1783. [CrossRef]
25. Schulz, A. Selektive Methanisierung von CO in Anwesenheit von CO_2 zur Reinigung von Wasserstoff unter den Bedingungen einer PEM-Brennstoffzelle. Ph.D. Thesis, University of Karlsruhe, Karlsruhe, Germany, 2005.
26. Thema, M.; Weidlich, T.; Hörl, M.; Bellack, A.; Mörs, F.; Hackl, F.; Kohlmayer, M.; Gleich, J.; Stabenau, C.; Trabold, T.; et al. Biological CO_2-Methanation: An Approach to Standardization. *Energies* **2019**, *12*, 1670. [CrossRef]
27. Sabatier, P.; Senderens, J.-B. Nouvelles synthèses du methane: New method for the synthesis of methane. *J. Chem. Soc.* **1902**, *82*, 333.
28. Chubb, T.A.; Nemecek, J.J.; Simmons, D.E. Application of chemical engineering to large scale solar energy. *Sol. Energy* **1978**, *20*, 219–224. [CrossRef]
29. McCrary, J.H.; McCrary, G.E.; Chubb, T.A.; Nemecek, J.J.; Simmons, D.E. An experimental study of the $CO_2 \times CH_4$ reforming-methanation cycle as a mechanism for converting and transporting solar energy. *Sol. Energy* **1982**, *29*, 141–151. [CrossRef]

30. Hasegawa, N.; Yoshida, T.; Tsuji, M.; Tamaura, Y. Integrated carbon recycling system for mitigation of CO_2 emissions utilizing degraded thermal energy. *Energy Convers. Manag.* **1996**, *37*, 1333–1338. [CrossRef]
31. Yoshida, T.; Tsuji, M.; Tamaura, Y.; Hurue, T.; Hayashida, T.; Ogawa, K. Carbon recycling system through methanation of CO_2 in flue gas in LNG power plant. *Energy Convers. Manag.* **1997**, *38*, S443–S448. [CrossRef]
32. Nitsch, J.; Pregger, T.; Sterner, M.; Wenzel, B. Langfristszenarien und Strategien für den Ausbau der Erneuerbaren Energien in Deutschland bei Berücksichtigung der Entwicklung in Europa und Global: Schlussbericht BMU–FKZ 03MAP146, Berlin. 2012. Available online: http://www.erneuerbare-energien.de/files/pdfs/allgemein/application/pdf/leitstudie2011_bf.pdf (accessed on 2 August 2021).
33. SRU. *Wege zur 100% Erneuerbaren Stromversorgung: Sondergutachten*; Erich Schmidt: Berlin, Germany, 2011; ISBN 978-3-503-13606-3.
34. UBA. *2050: 100% Erneuerbarer Strom: Energieziel 2050*; Umweltbundesamt: Dessau-Roßlau, Germany, 2010.
35. VDE ETG. *Energiespeicher für die Energiewende, Frankfurt am Main.* 2012. Available online: https://shop.vde.com/de/vde-studie-energiespeicher-für-die-energiewende-6 (accessed on 2 August 2021).
36. Edenhofer, O.; Pichs Madruga, R.; Sokona, Y. *Renewable Energy Sources and Climate Change Mitigation: Special Report of the Intergovernmental Panel on Climate Change*; Cambridge University Press: New York, NY, USA, 2012; ISBN 1107607108.
37. Thema, M.; Bauer, F.; Sterner, M. Power-to-Gas: Electrolysis and methanation status review. *Renew. Sustain. Energy Rev.* **2019**, *112*, 775–787. [CrossRef]
38. Specht, M.; Brellochs, J.; Frick, V.; Stürmer, B.; Zuberbühler, U. Technical Realization of Power-to-Gas Technology (P2G®): Production of Substitute Natural Gas by Catalytic Methanation of H_2/CO_2. In *Natural Gas and Renewable Methane for Powertrains: Future Strategies for a Climate-Neutral Mobility*, 1st ed.; van Basshuysen, R., Ed.; Springer: Berlin/Heidelberg, Germany, 2016; pp. 141–224, ISBN 978-3319232249.
39. Estermann, T. *Entwicklung eines Konzeptes zur Erweiterung der Thüga-Power-to-Gas-Demonstrationsanlage um eine Methanisierungseinheit: Development of a Concept to Expand the Power-to-Gas Demonstration Plant of the Thüga Group about a Methanation Unit. Bachelorarbeit*; OTH Regensburg: Regensburg, Germany, 2014.
40. Hentschel, J. Potenziale nachhaltiger Power-to-Gas Kraftstoffe aus Elektrizitätsüberschüssen im Jahr 2030. Ph.D. Thesis, Technical University of Berlin, Berlin, Germany, 2014.
41. Sterner, M.; Stadler, I. *Energiespeicher: Bedarf, Technologien, Integration*; Springer Vieweg: Berlin/Heidelberg, Germany, 2014; ISBN 978-3-642-37379-4.
42. Greenpeace Energy, eG. Rückenwind für die Energiewende: Der Gastarif Prowindgas. Available online: http://www.windgas.de (accessed on 2 August 2021).
43. Sterner, M.; Moser, A.; Thema, M.; Eckert, F.; Schäfer, A.; Drees, T.; Rehtanz, C.; Häger, U.; Kays, J.; Seack, A.; et al. *Stromspeicher in der Energiewende: Untersuchung zum Bedarf an neuen Stromspeichern in Deutschland für den Erzeugungsausgleich, Systemdienstleistungen und im Verteilnetz*; Agora-Speicherstudie 050/10-S-2014/DE: Berlin, Germany, 2014. [CrossRef]
44. Trost, T.; Horn, S.; Jentsch, M.; Sterner, M. Erneuerbares Methan: Analyse der CO_2-Potenziale für Power-to-Gas Anlagen in Deutschland. *Z Energ.* **2012**, *36*, 173–190. [CrossRef]
45. Jentsch, M. Potenziale von Power-to-Gas Energiespeichern: Modellbasierte Analyse des markt- und Netzseitigen Einsatzes im Zukünftigen Stromversorgungssystem. Ph.D. Thesis, University of Kassel, Kassel, Germany, 2014.

Article

Development of Stable Mixed Microbiota for High Yield Power to Methane Conversion

Márk Szuhaj [1], Roland Wirth [1], Zoltán Bagi [1], Gergely Maróti [2], Gábor Rákhely [1,3] and Kornél L. Kovács [1,4,*]

1. Department of Biotechnology, University of Szeged, 6726 Szeged, Hungary; szuhaj@bio.u-szeged.hu (M.S.); wirth.roland@brc.hu (R.W.); bagi.zoltan@brc.hu (Z.B.); rakhely.gabor@brc.hu (G.R.)
2. Institute of Plant Biology, Biological Research Centre, 6726 Szeged, Hungary; maroti.gergely@brc.hu
3. Institute of Biophysics, Biological Research Centre, 6726 Szeged, Hungary
4. Department of Oral Biology and Experimental Dentistry, University of Szeged, 6720 Szeged, Hungary
* Correspondence: kovacs.kornel@bio.u-szeged.hu

Citation: Szuhaj, M.; Wirth, R.; Bagi, Z.; Maróti, G.; Rákhely, G.; Kovács, K.L. Development of Stable Mixed Microbiota for High Yield Power to Methane Conversion. *Energies* **2021**, *14*, 7336. https://doi.org/10.3390/en14217336

Academic Editor: Jaakko Puhakka

Received: 21 September 2021
Accepted: 30 October 2021
Published: 4 November 2021

Publisher's Note: MDPI stays neutral with regard to jurisdictional claims in published maps and institutional affiliations.

Copyright: © 2021 by the authors. Licensee MDPI, Basel, Switzerland. This article is an open access article distributed under the terms and conditions of the Creative Commons Attribution (CC BY) license (https://creativecommons.org/licenses/by/4.0/).

Abstract: The performance of a mixed microbial community was tested in lab-scale power-to-methane reactors at 55 °C. The main aim was to uncover the responses of the community to starvation and stoichiometric H_2/CO_2 supply as the sole substrate. Fed-batch reactors were inoculated with the fermentation effluent of a thermophilic biogas plant. Various volumes of pure H_2/CO_2 gas mixtures were injected into the headspace daily and the process parameters were followed. Gas volumes and composition were measured by gas-chromatography, the headspace was replaced with N_2 prior to the daily H_2/CO_2 injection. Total DNA samples, collected at the beginning and end (day 71), were analyzed by metagenome sequencing. Low levels of H_2 triggered immediate CH_4 evolution utilizing CO_2/HCO_3^- dissolved in the fermentation effluent. Biomethanation continued when H_2/CO_2 was supplied. On the contrary, biomethane formation was inhibited at higher initial H_2 doses and concomitant acetate formation indicated homoacetogenesis. Biomethane production started upon daily delivery of stoichiometric H_2/CO_2. The fed-batch operational mode allowed high H_2 injection and consumption rates albeit intermittent operation conditions. Methane was enriched up to 95% CH_4 content and the H_2 consumption rate attained a remarkable 1000 mL·L^{-1}·d^{-1}. The microbial community spontaneously selected the genus *Methanothermobacter* in the enriched cultures.

Keywords: power-to-gas; thermophilic biogas; fed-batch reactor; *Methanothermobacter*; metagenome; starvation; H_2 and CO_2 conversion; methane; acetate

1. Introduction

The energy needs of civilized human lifestyle and the global population are increasing rapidly. The majority of this energy is provided currently from fossil energy carriers. Exploitation of fossil sources is associated with greenhouse gas (GHG) emission, which is the primary source of global climate change endangering the biosphere and overall quality of life on Earth. These are the driving forces for the increase of the contribution of renewable energy in the overall energy spectrum. Photovoltaics (PV), wind, hydro, and biomass are the major sources available [1,2]. The particularly rapid growth in energy production via PV and wind is much appreciated although the fluctuating nature of electricity generation using these forms of incoming solar energy presents additional challenges for the distribution and utilization systems [3,4]). Smart electricity grids and flexible storage technologies are being developed to balance the energy losses and grid imbalances due to the deranged production and utilization of electricity [5].

The mass-based energy content and carbon-free nature of hydrogen makes H_2 an excellent energy storage medium. H_2 can be produced in various ways [6], water electrolysis being the most commonly employed among them [7]. Conversion to a hydrogen economy [8–10] is an attractive scenario, which could help restore the rapidly deteriorating climate conditions. The primary hurdles to large scale H_2 use include underdeveloped storage and transport technologies, which are still costly and energy demanding [8,10].

Methane (CH_4) is also an excellent potential energy delivery material but it can contribute to GHG emission, due to its carbon content, unless generated via a renewable energy conversion process [11,12]. A significant advantage of CH_4 as an energy carrier is the efficient and advanced storage and transport pipeline system developed for natural gas, the fossil and thus less environmentally friendly form of CH_4 [13,14]. Biogas, a mixture of CH_4 (60–70), carbon dioxide (30–40%), and 1–2% of other gases [15], is generated during the anaerobic decomposition (AD) of biomass, a renewable form of stored solar energy continuously supplied on Earth via photosynthesis [16,17]. AD of biomass is carried out by a complex microbial community, biogas is formed in the last step of the multifarious biochemical process by methanogenic microbes. Based on their substrate preference, methanogens are classified in three groups, i.e., acetotrophic, hydrogenotrophic, and methylotrophic methanogens, all belonging in the phylum Euryarchaeota within the kingdom Archaea [18]. Hydrogenotrophic methanogens reduce carbon dioxide (CO_2) to CH_4 when the appropriate reducing power, i.e., H_2 or low redox potential electrons, is available. H_2 can be obtained from water electrolysis powered by renewable electricity, closing the circular character of this energy conversion process, called Power-to-Methane (P2M) [7,15,19,20]. P2M is accomplished either within the biogas producing AD reactor, i.e., in situ P2M, or in a separate reactor vessel, i.e., ex situ P2M, or in a combination thereof [15,19]. The advantages and disadvantages of the various reactor arrangements have been discussed extensively [15,21–23]. Methanogens, the key players in the P2M process can be employed in sterile, pure cultures [24–27] or in a mixed anaerobic microbial community [13,28–34]. An inexpensive, readily available source of the anaerobic methanogen community is the fermentation effluent of the biogas reactor itself, which is enriched in methanogens during the course of P2M [28,35].

A lab-scale proof of concept is presented in this study, in which the fermentation residue of an industrial thermophilic AD reactor is used to catalyze the P2M conversion of H_2 and CO_2 while the alterations of the microbial community under the selection pressure of the experimental conditions are established.

2. Materials and Methods

2.1. Fermentation System

The total volume of each batch reactor was 160 mL (Wheaton glass serum bottle, Z114014 Aldrich) and contained 40 mL fermentation effluent from the thermophilic industrial biogas plant Bátortrade Kft. Nyírbátor, Hungary. The main substrates at Bátortrade are animal waste (39.1%), manure (29.7%), agricultural waste (18.9%), and ensilaged green plant material (12.3%). The effluent, containing the "start" microbial community, was sieved on a 1 mm filter to remove the larger particles. In each set of experiments 3 control reactors, containing only the "start" inoculum, were included. The various reactors were operated in 3 parallel biological replicates. The reactors were sealed with butyl septa and aluminum crimps and the headspace was replaced with flushing by N_2 gas (Messer nitrogen 4.5) for 5 min. H_2 and CO_2 were injected manually and daily into the head-space with disposable plastic syringes. The reactors received varying volumes of daily H_2 doses, which were nominally 20, 40, and 60 mL of pure H_2 gas, respectively. The amount of the injected gas was verified by gas chromatography (GC) as corresponding to 18.0, 31.5 and 43.5 v/v% actual initial H_2 concentration in the head-space. The gas composition in the reactor head-space was determined daily by GC and after the measurements the reactors were degassed by purging with N_2 for 5 min and the internal pressure was adjusted to atmospheric level. The reactors were incubated in a thermostated rotary shaker at 55 °C.

2.2. Volatile Organic Acid Analysis

Samples for organic acid analysis were pretreated according to Szuhaj et al. [28].

The concentrations of volatile organic acids were measured with HPLC (Hitachi Chromaster) equipped with a refractive index detector Chromaster 5450. The separation was performed on an Agilent Hi-Plex H column. The temperature of the column and detector were 50 °C and 41 °C, respectively. The eluent was 0.02 M H_2SO_4 (0.6 mL·min^{-1}).

2.3. Gas Composition Analysis

The gas composition of the reactor headspace was measured every day by GC. The CH_4 and H_2 contents were determined with an Agilent 6890N GC (Agilent Technologies, Santa Clara, CA, USA) equipped with an HP Molesive 5 Å (30 m × 0.53 mm × 25 µm) column and a TCD detector. The temperature of the injector was 150 °C and application was made in split mode 1.1:1. The column temperature was maintained at 47 °C. The carrier gas was Linde HQ argon 5.0, with the flow rate set at 9.6 mL·min^{-1}.

The amount of CO_2 was determined with a Shimadzu GC 2010 (Shimadzu Corporation) equipped with a TCD detector and a HP PlotQ (30 m × 0.5 mm × 40 µm) column. The chromatograph was operated in split injection mode (rate 4:1). The temperature of the inlet was 200 °C. The column and the detector temperature were maintained at 90 °C and 150 °C, respectively. The carrier gas was Messer nitrogen 4.5 at 1.25 mL·min^{-1}. The samples were injected with the help of a gastight microsyringe (Hamilton). The conversion efficiency of H_2 to CH_4 was calculated by the modified theoretical equation [28,30,36].

$$\eta = \frac{(r_{CH_4A} - r_{CH_4B})}{(r_{H_2A} - r_{H_2D}) \times 4} \times 100$$

where "A" is the experimental reactor and
η = conversion efficiency of H_2 to CH_4 (%)
r_{CH_4A} = CH_4 production of reactor A (mL·L^{-1}·d^{-1})
r_{CH_4B} = CH_4 production of control reactor (mL·L^{-1}·d^{-1})
r_{H_2A} = injected amount of H_2 to reactor A (mL·L^{-1}·d^{-1})
r_{H_2D} = residual amount of H_2 in reactor A (mL·L^{-1}·d^{-1})

2.4. Determination of Fermentation Parameters

oDM: The organic dry matter content was quantified by drying the biomass at 105 °C overnight and weighing the residue, giving the dry mass content. Further heating of this residue at 550 °C provided the organic dry matter (oDM) content.

pH: The pH was measured with a Radelkis OP-211/2 equipped with an OP-0808P pH electrode immediately after the daily GC analysis.

2.5. Total DNA Isolation for Metagenomics

The composition of the microbial community was investigated twice during the experimental period from each reactor and controls, i.e., at the starting point (inoculation) and at the end of cultivation. For total community DNA isolation 2 mL samples were taken from each reactor. DNA extractions were carried out using the Zymo Research Fecal DNA kit (D6010, Zymo Research, Irvine, CA, USA). Bead beating was performed by Vortex Genie 2, bead size: 0.1 mm, beating time: 15 min, beating speed: max, and in other details the Zymo Research kit protocol was followed. The quantity of DNA was estimated using a NanoDrop ND-1000 spectrophotometer (NanoDrop Technologies, Wilmington, NC, USA) and a Qubit 2.0 Fluorometer (Life Technologies, Carlsbad, CA, USA). DNA purity was tested by agarose gel electrophoresis and on an Agilent 2200 TapeStation instrument (Agilent Technologies).

2.6. Shotgun Sequencing

The recommendations of the Ion Torrent PGM™ sequencing platform were closely followed (Life Technologies, Carlsbad, CA, USA). The preparation of sample libraries was done according to the Life Technologies IonXpress fragment plus library protocol (4471269). Ion device library quantitation kit (4468802) and Step One Real Time PCR

(Applied Biosystems) were used to quantify the samples. The Ion PGM Template OT2 200 kit (4480974) was used with OneTouch 2 and Ion OneTouch ES devices. The barcoding was done by IonXpress barcode kit (4471250). Sequencing was performed with Ion PGM 200 Sequencing kit (4474004) on Ion Torrent PGM 316 chip.

Raw sequences are available on NCBI Sequence Read Archive (SRA) under the submission number: PRJNA625695.

2.7. Raw Sequence Filtering

The Galaxy Europe server was employed to pre-process the raw sequences (i.e., sequence filtering, mapping, quality checking) [37]. Low-quality reads were filtered by Prinseq [38] (min. length: 60; min. score: 15; quality score threshold to trim positions: 20; sliding window used to calculated quality score:1). Filtered sequences were checked with FastQC.

2.8. Read-Based Metagenome Data Processing and Statistical Analysis

The filtered sequences were further analyzed by Kaiju, applying the default greedy run mode on Progenomes2 database [39,40]. MEGAN6 was used to investigate microbial communities and export data for statistical calculation. The results were plotted with iTOL (Interactive Tree of Life) [41]. The microbial changes of the communities were estimated as the log2 fold changes (log2FC):

$$\log 2FC = \log 2 \left(\frac{\text{abundance X}}{\text{abundance START/CONTROL}} \right)$$

3. Results

3.1. Methanogenesis by H_2 and $H_2 + CO_2$

The mixed AD community was first supplied with various amounts of H_2 in order to eliminate the dissolved CO_2/HCO_3^- in the AD fermentation effluent (Figure 1).

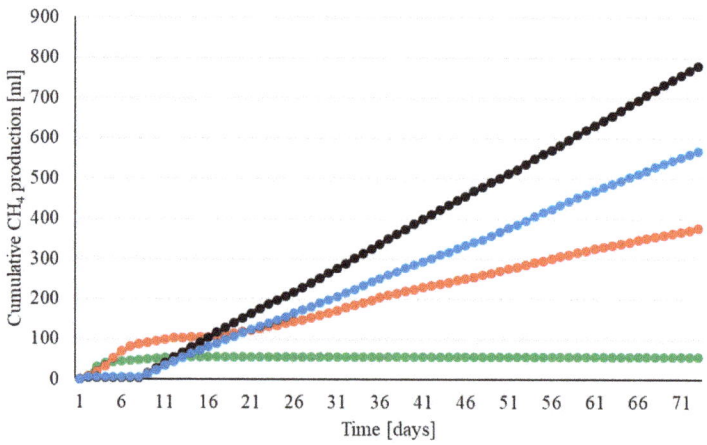

Figure 1. Cumulative biomethane productions in time from 20 mL nominal H_2 volume (=18.0 $v/v\%$ H_2, red curve), 40 mL nominal H_2 volume (=31.5 $v/v\%$ H_2, blue curve) and 60 mL nominal H_2 volume (= 43.5 $v/v\%$ H_2, black curve) and concomitant stoichiometric CO_2. The control, i.e., only daily N_2 gas replacement of the head-space, is shown in green. The symbol sizes indicate the error of measurement.

The control samples (green curve) evolved a small amount of residual CH_4 during day 1–2, but CH_4 generation ceased afterwards indicating the cessation of biogas formation due to the preceding depletion of degradable organic substrates. The reactors received varying volumes of daily H_2 doses, which were nominally 20, 40, and 60 mL of pure H_2 gas, shown with red, blue and black curves, respectively, corresponded to 18.0, 31.5, and 43.5 $v/v\%$ actual initial H_2 concentration in the head-space. The daily H_2 doses were completely consumed within 24 h in all reactors. The microbial community quickly consumed the dissolved CO_2/HCO_3^- as well, indicating high biological activity. The reactors receiving 18 $v/v\%$ of H_2 started to produce CH_4 intensively, implying sufficient level of a hydrogenotrophic methanogen activity for the bioconversion reaction. After about a week of daily H_2 feeding of the reactors the CH_4 evolution began to cease and the cumulative CH_4 production curve levelled off. By this time the CO_2 completely disappeared from the headspace of the reactors (Figure 2). A combination of these observations was indicative of methanogenesis limitation as a consequence of CO_2/HCO_3^- depletion by hydrogenotrophic methanogenesis. The situation was remedied by the injection of 6.5 $v/v\%$ CO_2 together with the daily H_2 dosage (Figure 2) on days 8–14 and 22 as indicated by the arrows.

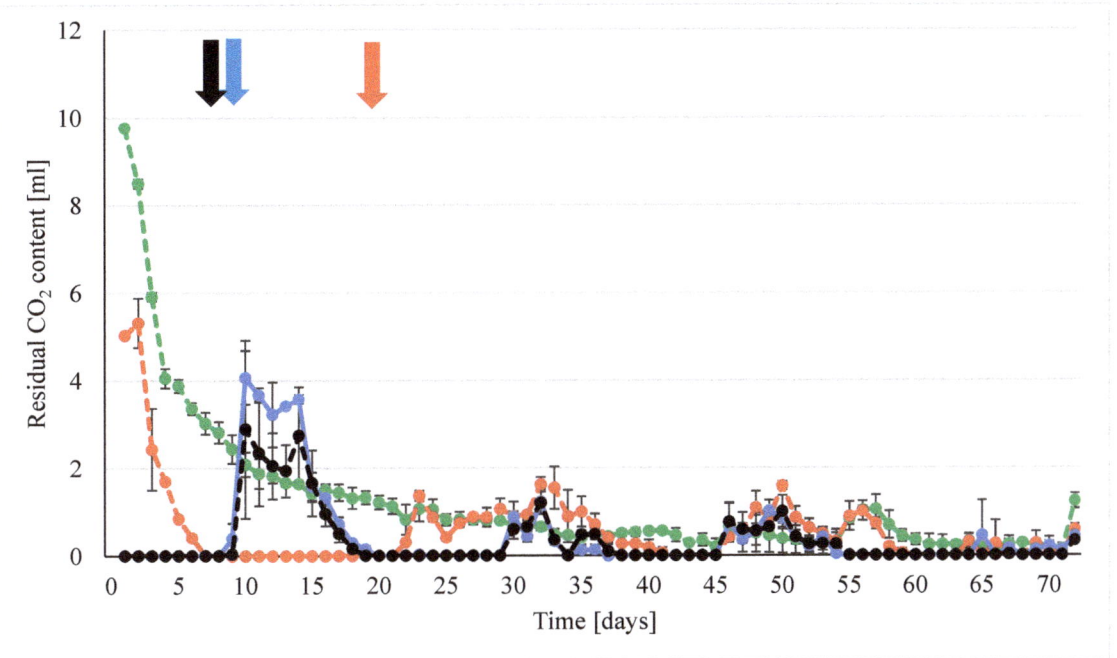

Figure 2. The residual CO_2 levels measured in the reactors' head-space. Color codes as in Figure 1: green = control reactors, red = low H_2 daily dosage (20 mL H_2 nominal volume), blue = medium H_2 daily dosage (40 mL H_2 nominal volume), black = high H_2 daily dosage (60 mL H_2 nominal volume). The arrows indicate the time points when excess CO_2 was delivered in order to remedy system imbalance.

This effectively restored CH_4 production, although the excess CO_2 supply resulted in a transient accumulation of CO_2 in the headspace. Henceforth, a 4:1 volumetric mixture of H_2 and CO_2 was injected daily into the head-space of the reactors. Steady CH_4 production, without detectable H_2 or CO_2, was maintained throughout the rest of the 71-day long experiment, demonstrating sustainable bioconversion of H_2/CO_2 to CH_4. The alterations in the pH of the reactor content reflected and corroborated the postulated sequence of events. Following week 2, a significant pH elevation was noted in the reactors receiving 18 $v/v\%$ of H_2, which quickly reached an alarming level above pH = 9 (Figure 3) on weeks 2 and 3. The injection of CO_2 slowly returned the pH level to normal, corroborating the essential role of CO_2/HCO_3^- in maintaining and regulating the buffering capacity in these systems.

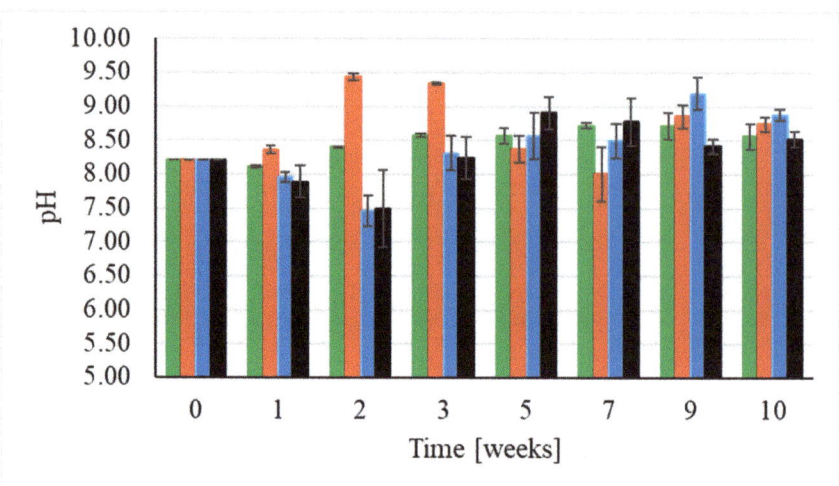

Figure 3. The alterations in pH values of the aqueous phase of the reactors: green = control reactors, red = 20 mL H_2 nominal volume daily dosages, blue = 40 mL H_2 nominal volume daily dosages, black = 60 mL H_2 nominal volume daily dosages and concomitant stoichiometric CO_2.

The reactors receiving higher H_2 doses (31.5 or 43.5 $v/v\%$) showed a substantially distinct behavior. In these reactors CH_4 evolution did not commence upon H_2 addition and CO_2 was not detectable in the headspace even after the first day. Nevertheless, the injected H_2 was consumed completely by the microbial community. Pursuing possible bioconversion product(s) other than CH_4 revealed that the microbes utilized the H_2 and CO_2 in the reactors for syntrophic acetate production via homoacetogenesis [42]. Accumulation of considerable amounts of acetate (Figure 4) indicated the predominance of this pathway in these reactors. In line with this mechanism was the substantial pH drop on week 2 (Figure 3). Overdosed (9.9 and 13.9 $v/v\%$) injection of CO_2 (Figure 2) successfully balanced the pH back to near normal level for methanogenesis in these reactors. Accordingly, steady CH_4 production started (days 8–10) and daily stoichiometric gas delivery of H_2 and CO_2, drifted the system away from volatile fatty acid (VFA) biosynthesis to hydrogenotrophic methanogenesis (Figures 1–3). All reactors were eventually stabilized in the biomethane production mode and maintained their stable operation for the rest of the experimental period. Acetate and other VFAs were barely detectable in these reactors (Figure 4).

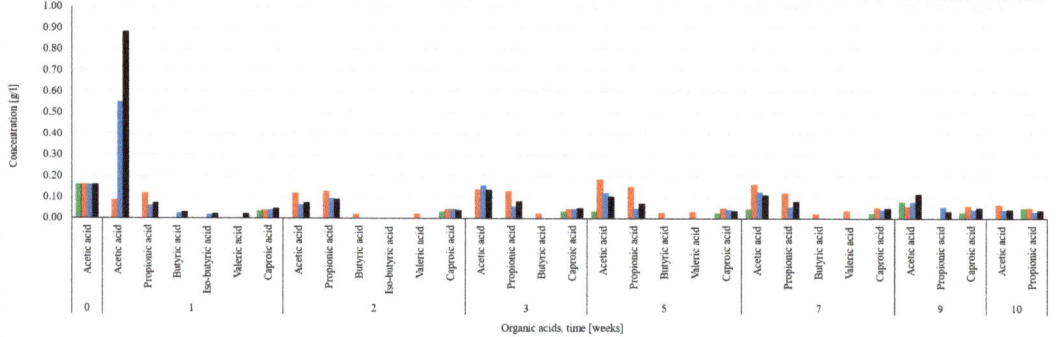

Figure 4. The volatile fatty acids (VFA) productions of the P2M reactors: green = control reactors, red = 20 mL H_2 nominal volume daily dosages, blue = 40 mL H_2 nominal volume daily dosages, black = 60 mL H_2 nominal volume daily dosages and concomitant stoichiometric CO_2.

3.2. Metagenomic Analyses

The microbial community of the thermophilic digestate was diverse, although the majority of the identified genera were present in low relative abundance (<1%) (Figure 5). Feeding the community with only H_2/CO_2 daily, acted as strong selection pressure on the community by the end of the 71-day long P2M experiment. In spite of the apparent high initial microbial diversity, a limited number of taxa survived the P2M experimental conditions (Table 1) leading to the substantial enrichment of the successful survivors.

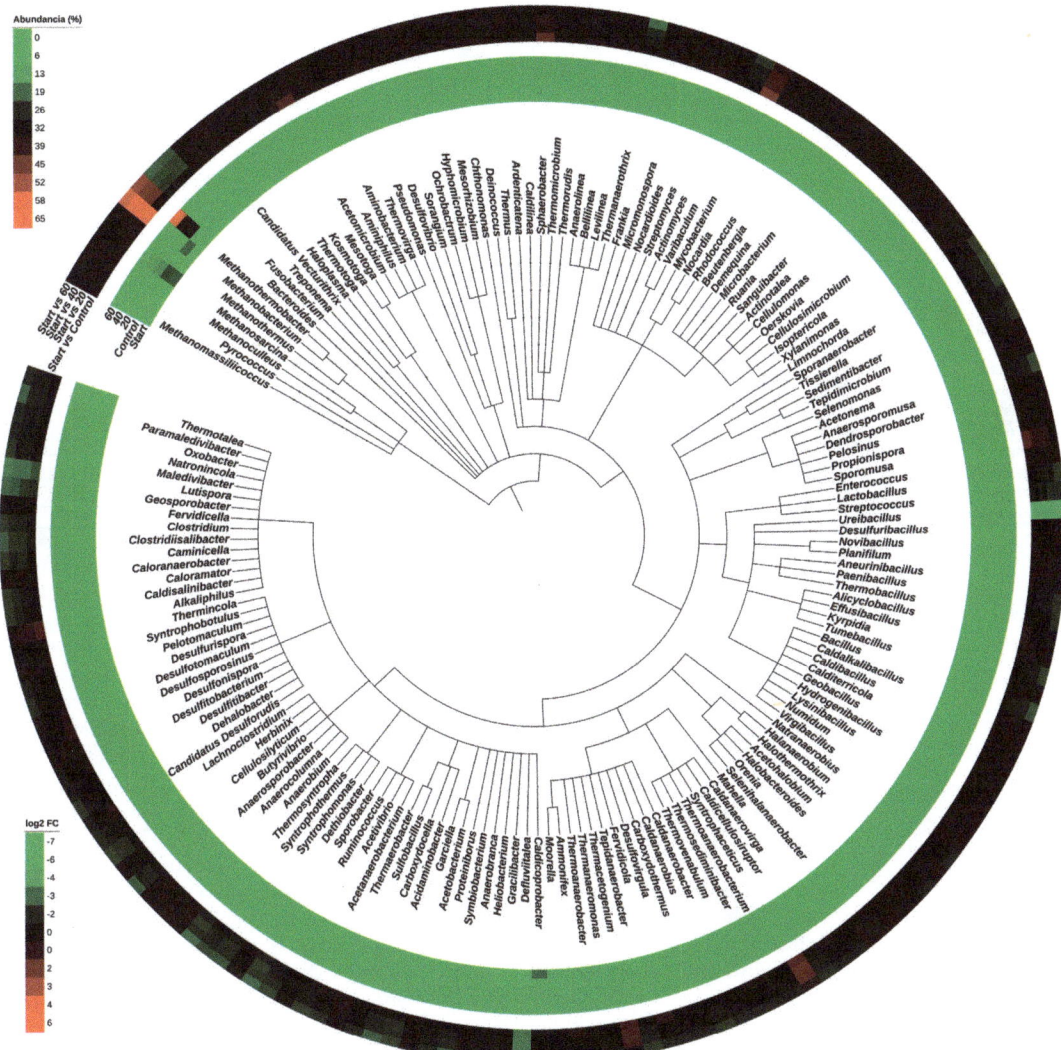

Figure 5. The microbial diversity map is plotted in the central taxonomic tree and the names of genera corresponding to the branches. The average relative abundances of the taxa in the various reactors are shown in the set of the outer 5 rings: starting community; control (= no H_2 feeding) community; 20, 40, and 60 mL H_2 nominal volume daily dosages, respectively; color scale in the upper left corner. Set of 4 outermost rings: comparison of the logarithmic fold change (log2FC) values between the start community and the control, 20, 40, and 60 mL H_2 nominal volume daily dosages, respectively, and concomitant stoichiometric CO_2 (color scale in the lower left corner).

Table 1. The changes in relative abundances of the 23 most abundant microbial taxa upon various treatments. START: composition of the initial thermophilic AD effluent; Control: only daily N_2 gas replacement of the head-space; 20 mL H_2 reactor: 20 mL nominal H_2 volume (=18.0 $v/v\%$ H_2); 40 mL H_2 reactor: 40 mL nominal H_2 volume (=31.5 $v/v\%$ H_2); 60 mL H_2 reactor: 60 mL nominal H_2 volume (=43.5 $v/v\%$ H_2) and concomitant stoichiometric CO_2. Color codes in the cells of taxon names are as follows: red—Archaea genera; white—the abundance was not affected by H_2/CO_2 addition; yellow—the abundance of genera decreased at all H_2/CO_2 concentrations as a result of starvation stress; green—genera responding positively to starvation and to moderate H_2/CO_2 daily dosage. Red background color in the log2FC cells indicates enrichment, blue background color denotes diminishing tendency for the given taxon. The most striking changes are the boxed in heavy borderlines.

Taxon	START	Control	20 mL H_2 Reactor	40 mL H_2 Reactor	60 mL H_2 Reactor
Methanosarcina	21.69	22.80	15.24	13.70	9.35
Caldicoprobacter	19.89	7.12	12.32	9.90	4.20
Ureibacillus	9.10	0.05	0.19	0.12	0.05
Methanothermobacter	8.37	6.60	29.47	37.53	65.26
Clostridium	4.52	1.48	1.02	1.01	0.78
Lutispora	2.28	0.29	0.17	0.16	0.13
Desulfotomaculum	1.80	2.18	1.41	1.30	0.78
Moorella	1.53	1.26	0.85	0.73	0.43
Herbinix	1.35	0.50	0.27	0.28	0.22
Lysinibacillus	1.33	0.10	0.10	0.10	0.10
Limnochorda	1.31	1.09	0.63	0.50	0.23
Acetomicrobium	1.28	3.11	1.54	1.32	0.78
Bacillus	1.23	0.50	0.35	0.31	0.16
Paenibacillus	1.16	0.60	0.40	0.39	0.23
Syntrophomonas	1.03	0.36	0.22	0.21	0.13
Tepidanaerobacter	1.01	0.65	0.34	0.25	0.17
Syntrophaceticus	0.57	1.48	1.79	1.55	0.76
Sphaerobacter	0.57	1.60	0.67	0.76	0.36
Thermacetogenium	0.55	1.11	1.24	1.18	0.51
Mycobacterium	0.37	1.69	0.80	0.78	0.55
Methanobacterium	0.28	19.99	10.99	9.30	5.72
Actinotalea	0.10	5.09	5.32	5.39	2.10
Cellulomonas	0.10	2.44	2.65	2.66	1.05

Within the kingdom *Bacteria* the genus *Caldicoprobacter* turned out to be the most abundant (19.9%) in the starting community, i.e., in the fermentation effluent of the thermophilic AD plant. These hydrolyzing bacteria belong in the order *Clostridia* and class *Firmicutes* and degrade various carbohydrates, e.g., arabinose, xylose, ribose, fructose [43], and proteins via their active serine protease [44]. The second most abundant genus in the kingdom Bacteria was *Ureibacillus*. These bacteria can also carry out a number of heterotrophic decomposition pathways [45–47]. Additional predominating members of the thermophilic anaerobic community were the genera *Clostridium* and *Lutispora* with relative abundances of 4.5 and 2.3%, respectively. These genera are routinely found in biogas communities [48,49], together with the less abundant genera *Desulfotomaculum* (1.8%) and *Moorella* (1.5%) (Table 1).

Members of the genus *Methanosarcina* were initially the predominant ones among methanogens with relative abundance of 21.7%. *Methanosarcina* is the only known genus, which is able to carry out all three methanogenic pathways, i.e., acetoclastic, hydrogenotrophic, and methylotrophic CH_4 biosynthesis. This versatile capability makes them the most frequently detected methanogens in many biogas producing systems [7,15]. In line with the metabolic versatility, they showed excellent survival competence upon starvation, the relative abundance in the control reactor was 22.8%. Next in abundance among the initial methanogens was the genus *Methanothermobacter* (8.4%), a typically hydrogenotrophic methanogen [50]. They endured the starvation just as well as the genus *Methanosarcina*, and eventually became the most predominant methanogens in all reactors fed with H_2/CO_2, greatly outnumbering the other two methanogenic genera and therefore substantially contributing to the P2M conversion. It is noteworthy, that the relative abundance of the genus *Methanothermobacter* increased with the daily H_2 doses injected into the reactors while both *Methanosarcina* and *Methanobacterium* appeared to respond on the contrary, i.e., their relative abundances apparently decreased at elevated H_2 addition (Table 1). Since these abundance values are relative ones, they indicate the level of competition for H_2 among the methanogens rather than the absolute number or survival vigor of these Archaea.

The overall rearrangement of the microbial community as a result of H_2/CO_2 feeding may not look spectacular at first glance (Figure 5, outermost rings). This is due to the fact that sequencing of the samples allowed the identification of numerous taxa present in very low abundance, i.e., <0.1% in the microbial communities. The changes in the scarcely present microbes upon various treatments are difficult to assess, therefore these were not considered in the comparative analyses. Important changes were recognized upon a closer look (Table 1) of the 23 most abundant taxa present in the starting microbial community. Although fresh organic substrate was not delivered into the reactors, some of the heterotrophic *Bacteria* managed to survive and flourish in spite of the lack of added organic substrates for their heterotrophic growth.

A marked reorganization of the microbial community took place, when the microbes were subjected to starvation, i.e., neither external organic substrate nor H_2/CO_2 was available to support their life. Comparison of the "start", i.e., thermophilic AD effluent, microbial community with the "control", i.e., thermophilic AD effluent incubated at 55 °C, with daily replacement of the headspace with N_2 gas, clearly indicated a fight for survival within the community (Table 1, columns 2 and 3).

The regulatory effects of H_2 and/or H_2/CO_2, together with the lack of added organic substrates, manifested themselves in the genera *Ureibacillus* (log2FC= −7.6), *Lutispora* (log2FC= −3.9), *Herbinix* (log2FC= 2.4), *Clostridium* (log2FC= −2.3), *Bacillus* (log2FC= −1.3), *Tepidanaerobacter* (log2FC= −2.0) among Bacteria. *Ureibacilli* have been found frequently in thermophilic aerobic poultry waste treatment sites [51]. Similarly, the genera *Lysinobacillus* and *Paenibacillus* are typical components of the poultry manure microbiota [52,53]. These apparently "outlier" bacteria (taxon names are highlighted in yellow in Table 1) could have therefore infiltrated the thermophilic AD community from the AD substrate, which contained poultry meat processing waste (Figure 6).

The declining representation of homoacetogens, e.g., genus *Syntrophomonas* (log2FC= −2.5) [54], indicated a shift from syntrophic acetate oxidation (SAO) to hydrogenotrophic methanogenesis in the reactors fed with increasing H_2/CO_2 loading. In contrast, the SAO genus *Actinotalea* acclimated excellently to the starvation condition and subsequently to the P2M conditions. Similarly, other SAOB genera, such as *Syntrophaceticus* and *Thermoacetogenium* [55] emerged substantially from the diverse group of low abundant taxa in the starting community upon starvation and remained stable members of the community, although virtually unaffected by the amount of daily H_2 dosage (taxon names are highlighted in green in Table 1). Others (taxon names are highlighted in white in Table 1) remained unaffected by the presence of H_2/CO_2 in their environment.

Interestingly, the methanogens (taxon names are highlighted in red in Table 1) responded differently to the changing environment. The genus *Methanosarcina*, a predominant methanogen in the starting community, was not affected by the starvation, but became slightly inhibited by the daily dosage of H_2/CO_2 (log2FC = −0.8), although it should be noted again that only relative abundance values were compared in this study. Increasing the H_2 supply triggered an apparent drop in the number of *Methanobacteria*, but their average abundance was at the respectable log2FC = 4.9. The hydrogenotrophic methanogen Archaea (HMA), belonging in the genus *Methanothermobacter* became the absolute predominant taxon, its relative abundance increased from 8.4% in the starting community up to 65.3% in the reactors receiving 43.5 v/v% H_2 + 10.9 v/v% CO_2 every day (Figure 6).

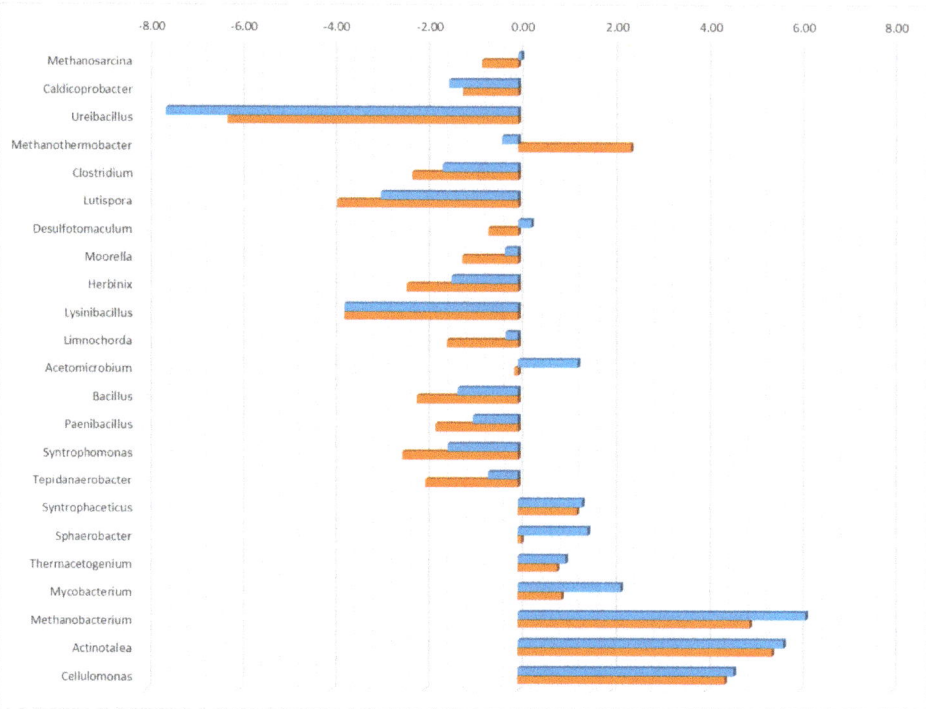

Figure 6. The logarithmic fold change (log2FC) abundance alterations in the most abundant 23 genera at the end of the experiment. The control vs. starting community log2FC values are plotted in green columns, and the log2FC between the control vs. the average of H_2/CO_2-fed reactors are shown in red columns.

4. Discussion

Signatories of the Paris agreement [56] pledged to exercise joint efforts and limit the global climate change to below 2 °C, a challenging undertaking for mankind. Among the measures to be implemented, the expanded use of renewable energy carriers to replace the currently predominating fossil ones has high priority. There are a number of technical tasks to solve and obstacles to overcome along this road, but photovoltaics, wind, and hydro power technologies have already gained momentum and increased appreciably the contribution to the conversion of the energy palette. These technologies produce renewable electricity often in a fluctuating and unpredictable fashion, depending on the environmental conditions. The electricity grids experience difficulties in handling this irregular input without major losses. The power-to-gas (P2G) or power-to-methane (P2M) technology offers a solution by converting the excess renewable electricity to H_2 or CH_4 via P2G or P2M, respectively [57]. H_2 is a carbon-free energy carrier for the future, CH_4 seems to be an excellent second choice and an easier target due to the existing elaborated transport and storage networks for "natural gas", which is essentially fossil CH_4.

Several approaches have been proposed to realize P2M [15] using biological systems as catalyst of the CO_2 reduction to CH_4. They all employ hydrogenotrophic methanogenic Archaea (HMA), which carry out the required biochemical reaction in their natural environment to sustain their life [58], within the biogas producing anaerobic microbial communities [59]. HMAs can accomplish CO_2 reduction both in microbiologically pure cultures and as parts of a larger microbial community. In a mixed community the provision of only H_2 or H_2/CO_2 acts as selection pressure in the process. Under mesophilic conditions [28,35] the reactors fed solely with H_2, consumed the dissolved CO_2/HCO_3^- and CH_4 evolution ceased after 26–28 days. Correcting the conditions by daily injection of stoichiometric H_2/CO_2 in the head-space restored the P2M process, which could be maintained for an extended period of time. The daily doses of H_2/CO_2 were completely consumed within 16–20 h.

In the present study the same overall workflow was tested under thermophilic (55 °C) conditions. Elevation of the reaction temperature accelerated the performance of the HMAs in the mixed microbial community. The reactors supplied with 31.5 or 43.5% H_2 did not evolve CH_4 (Figure 1), but converted the H_2/CO_2 to primarily acetate (Figure 4), which suggested a homoacetogenic reaction pathway for H_2 metabolism instead of the expected CH_4 evolution. Apparently, the dissolved CO_2/HCO_3^- was used as carbon source in the acetate biosynthesis as indicated by both the VFA profile and the increased pH by the end of day 7. Similar events were observed only after 3–4 weeks of daily H_2 injection under mesophilic conditions [28].

Distinct behavior was observed at low daily H_2 dosage, i.e., 18% of H_2 in the head-space at the time of gas injection. The H_2 also wholly disappeared from the head-space within 24 h and CH_4 evolution commenced. Accumulation of VFAs and concomitant elevation of pH was not observed in these reactors indicating that the P2M reaction took place already during this first period of the experiment. By the end of week 2 probably most of the dissolved CO_2/HCO_3^- was converted to CH_4. Elevation of the pH was the first warning sign of system imbalance, which required action to restore stable CH_4 production. Addition of stoichiometric daily H_2/CO_2 gas mixture resolved the problem.

In summary, the entire experimental timeline could be divided clearly into two phases. During the first period, the system responded quickly but in a distinct manner depending on the supplied H_2 level and the dissolved CO_2/HCO_3^-. When only low H_2 volumes were injected into the reactors, i.e., 18 v/v%, CH_4 evolution took place as expected. In contrast, addition of higher dosage of H_2 inhibited CH_4 formation completely from the start. Supplementation of a stoichiometric mixture of H_2/CO_2 yielded 87.5–95.5% CH_4 content during the second, stabilized P2M generation period in all reactors.

A comparison of the performance using a fed-batch and "H_2 flow-through" reactor arrangement reveals the advantages of this P2M approach relative to the widely studied flow-through CSTR (continuous stirred-tank reactor) reactor configuration [29,60,61] (Table 2). The fed-batch reactors upgraded the gas mixture to 95% bioCH_4 purity, which is close to the methane content required for direct injection into the natural gas grid. The yield of CH_4 production from the injected H_2 doubled in comparison to the values reported in [29,62]. The H_2 injection rate reached 1300 mL·L^{-1}·d^{-1} in the reactors receiving the highest daily H_2 dose and practically all the injected H_2 was consumed by the microbial community for CH_4 production. Moreover, the community apparently did not demand additional nutrients and/or minerals to sustain the biological activity of the enriched community as opposed to the systems utilizing a single HMA strain [22]. Taking these advantages together (Table 2), the fed-batch reactor arrangement and the intermittent gas feeding are recommended to become a novel, efficient P2M strategy as opposed to the flow-through P2M reactors. The increased residence time of H_2/CO_2 in contact with the aqueous bulk, containing the key living and actively functioning microbial catalysts, offers improved reactor performance although the technology is still far from being optimized. The costs of installation and operation of a fed-batch type P2M reactor would be substantially lower than that of a more sophisticated flow-through in situ or ex situ reactors equipped with online GC monitoring, fine-tuned process control, gas recirculation and management systems [29,60,61].

The results corroborated the existence of a delicate balance between homoacetogenesis and methanogenesis, which is closely regulated by the H_2 level within the system [20,62,63]. The time resolution of sampling, analytical measurements, and the complexity of the biochemical reactions in the system did not allow calculation of a precise mass balance in this system. Nevertheless, the data are compatible with the assumption that no additional metabolic pathway contributed considerably to the outcome of the P2M reaction by the mixed anaerobic thermophilic microbial communities. Detailed mapping of the events in time within the P2M reactors is needed as part of future research efforts to determine the boundaries and precise operational parameters, kinetics of dissolution of the gases and the development of H_2 concentration gradients in the aqueous medium, etc. This is important for planning and management of the large scale P2M facilities based on mixed microbial systems, which should be competitive compared to the microbiologically well-defined P2M technologies using a pure HMA culture, particularly if the savings in operational costs are taken into account [7,15,16].

Two effects causing selection pressure could be distinguished in these experiments. First, there was an alteration in the composition of the community composition as the result of starvation, i.e., incubation of the community under anaerobic condition for an extended period of time. Second and added to the starvation stress, the community responded to the addition of varying amounts of H_2/CO_2 so that certain genera were enriched, while others tended to diminish.

Among Bacteria, members of the genera *Actinotalea* (log2FC = 5.1) and *Cellulomonas* (log2FC = 2.4) were apparently the ones to respond successfully to the selection pressure brought about by starvation. *Actinotalea* are facultative anaerobes [64,65] and thus carry out versatile metabolic pathways, whereas members of the genus *Cellulomonas* are superior in polysaccharide decomposition [66]. Since no organic material was supplied to support their heterotrophic growth, they could obtain organic substrate from the deceased members of the bacterial community. It is noteworthy that some genera present in low abundance (≤0.5%) in the inoculum, e.g., *Syntrophaceticus*, *Sphaerobacter*, *Thermacetogenium*, and *Mycobacterium* also managed to avoid extinction under the starvation conditions [67–70]. Some of these and the genus *Acetomicrobium* exhibited similar behavior and are suspected or verified SAOBs [71].

The methanogenic genus, *Methanosarcina*, apparently did not suffer much from starvation stress (Table 1) and maintained its predominance in the AD community [59,72,73]. The abundance of the genus *Methanothermobacter* was also largely unaffected by the starvation stress. The hydrogenotrophic methanogen genus *Methanobacterium* took a surprisingly pronounced advantage under starvation condition, similarly to a recent finding at mesophilic temperature [49]. This taxon increased its relative abundance from a meager 0.3% to 20% under starvation conditions. Future metatranscriptomic studies should reveal the metabolic changes responsible for this outstanding behavior.

The changes in the abundances of the various taxa upon supplying the starving reactors with various daily dosages of H_2/CO_2 is of particular interest with regard to the development and microbiological management of a stable and efficient microbial P2M community. The differences in abundances between the microbial communities subjected to starvation stress and equally starved and H_2/CO_2-fed communities are expressed at logarithmic scale (log2 fold change, Figure 6). The aim of this comparison was to separate the alterations in community composition due to H_2/CO_2 supply and starvation. It should therefore be noted that the lengths of the horizontal columns do not indicate absolute or relative abundances of any given taxon, these columns show only the differences between starving and H_2/CO_2-fed reactors.

Table 2. Comparisons of process parameters determined in this study to relevant reported performance characteristics. "n.d." stands for "not determined". The "A", "B" and "C" reactor groups indicate the low, medium and high H_2 daily dosage conditions as in Figure 1.

	Bassani et al. (2015)		Corbellini et al. (2018)			Mulat et al. (2017)			Present Work		
	Control	H_2 Added	Control	CSTR	UASB	Control	H_2 Added	Control	"A" Reactor After stabilization	"B" Reactor After stabilization	"C" Reactor After stabilization
Biogas composition (%)											
CH_4	67.1 ± 0.8	85.1 ± 3.7	69.2 ± 1	86.4 ± 1.0	91.0 ± 2.0	65.2 ± 9.8	89.4 ± 0.4	38.57 ± 1.96	87.48 ± 5.86	95.48 ± 5.01	94.13 ± 4.90
CO_2	32.9 ± 0.9	6.6 ± 0.9	30.7 ± 1	10.7 ± 3.6	7.0 ± 1.0	34.8 ± 9.8	10.5 ± 0.4	61.43 ± 8.51	6.69 ± 2.31	1.56 ± 1.00	1.04 ± 0.56
H_2	0.0	8.3 ± 3.6	0.0	3.5 ± 1.5	2.0 ± 1.0	0.0	0.0	0.0	2.99 ± 1.61	1.90 ± 0.90	3.49 ± 2.50
Gas production (mL·L^{-1}·d^{-1})											
CH_4	247 ± 27	359 ± 20	n.d.	n.d.	352.53 ± 53	224	267	19.8 ± 0.4	131.2 ± 14.4	197.3 ± 4.4	270.7 ± 7.4
CH_4 from H_2	0	112.0	0.0	n.d.	n.d.	0.0	127.0	0.0	111.4	177.5	250.9
CO_2	121 ± 15	28 ± 5	n.d.	n.d.	28.03 ± 4.7	119.6	31.4	31.0 ± 3.3	16.1 ± 0.7	11.1 ± 0.8	8.5 ± 1.8
H_2 injection rate (mL·L^{-1}·d^{-1})	0.0	510 ± 32	0.0	550	n.d.	0.0	507 ± 30	0.0	540.3 ± 71.6	945 ± 51.3	1306 ± 45.5
H_2 consumption (mL·L^{-1}·d^{-1})	0.0	470 ± 35	0.0	n.d.	n.d.	0.0	507 ± 30	0.0	509.00 ± 1.67	854.9 ± 3.48	1168.41 ± 4.98
H_2 consumption (%)	0.0	92.20	0.0	n.d.	n.d.	0.0	100.00	0.0	99.05	99.40	98.90
pH	7.82 ± 0.16	8.49 ± 0.04	n.d.	8.6 ± 0.0	8.1 ± 0	7.49 ± 0.17	7.00–8.00	8.57 ± 0.19	8.75 ± 0.11	8.88 ± 0.09	8.52 ± 0.11
Organic acids (g·L^{-1})	1.18 ± 0.84	0.38 ± 0.07	0.2	2.7	0.1	n.d.	n.d.	n.d.	n.d.	n.d.	n.d.
Acetate	n.d.	n.d.	n.d.	n.d.	n.d.	n.d.	n.d.	0.00	0.06	0.04	0.04
Propionate	n.d.	n.d.	n.d.	n.d.	n.d.	n.d.	n.d.	0.05	0.05	0.03	0.04

Practically none of the most abundant 20 bacterial genera responded with elevated growth (Figure 6) to the daily H_2/CO_2 dosage into the headspace of the reactors although the microbes in the fed-batch reactors consumed the injected H_2 completely within 16–24 h at all three H_2 concentrations. The genus *Ureibacillus* did not follow the general trend, but its representation was severely decimated from 9.1% to 0.05% upon starvation (Table 1), therefore it became negligible and could not interfere with the life of the community. Moderate apparent inhibition by H_2/CO_2 was observed in the case of all bacterial genera (Figure 6). This was likely due to the fact that relative abundance values were used in the calculation of log2FC and the predominance of the hydrogenotrophic genus *Methanothermobacter*, increased substantially in the H_2/CO_2-fed reactors.

Nevertheless, the presence of SAOBs among the most abundant 20 bacterial strains is noteworthy and the suspected SAOB-methanogen syntrophic contribution to the improvement of the P2M conversion demands further detailed study involving metatranscriptomics.

5. Conclusions

There are two "take home" messages from the studies reported in this paper. First, the fed-batch reactor configuration should be considered as an alternative to the widely used flow-through arrangement. The flow-through reactors are fine-tuned to minimize H_2 loss in the effluent gas, therefore very low H_2 injection rates are employed, which limits the attainable CH_4 production rates. In addition, the infrastructure and delicate process control makes these approaches costly. Although a fed-batch P2M reactor works intermittently, the added benefits, i.e., high H_2 loading rate, complete conversion of H_2/CO_2 to CH_4 and low operation costs, can make this approach appealing for future scale-up development.

Second, we demonstrated that the genus *Methanothermobacter* is enriched as the sole predominant methanogenic taxon under the selection pressure of the P2M conditions. Consequently, a mixed microbial community from a thermophilic AD plant can simply be used as catalyst in the P2M reactors after a few days/weeks of enrichment period and maintenance of microbiologically pure conditions; addition of expensive complex medium and micro nutrients are not necessary. In so doing, the costs of the P2M operation at an industrial scale can be reduced substantially.

In future studies the economic analysis and the larger scale testing of the proposed novel P2M process are required.

Author Contributions: Conceptualization, K.L.K. and Z.B.; methodology, M.S. and R.W.; formal analysis and experimental investigation, M.S., G.M., and R.W.; resources, K.L.K. and Z.B.; data curation, R.W.; writing—original draft preparation, M.S. and K.L.K.; writing—review and editing, K.L.K. and Z.B.; supervision, G.R.; project administration, Z.B.; funding acquisition, K.L.K., G.R., R.W., G.M. All authors have read and agreed to the published version of the manuscript.

Funding: This work was supported by the European Regional Development Fund to projects IN-NOV-397-13/PALY-2020 (PI: G.R), 2020-1.1.2-PIACI KFI (PI: Z.B.), and by the bilateral scientific collaboration agreement 2019-2.1.13-TÉT_IN-2020-00016 (PI: K.L.K.). R.W. (PD132145), Z.B. (FK 123902) and G.M. (FK123899) received support from the National Research, Development and Innovation Office (NKFIH), Hungary. G.M. was also supported by the Lendület-Programme of the Hungarian Academy of Sciences (LP2020-5/2020).

Institutional Review Board Statement: Not applicable.

Informed Consent Statement: Not applicable.

Data Availability Statement: The raw sequence data obtained in this project have been uploaded on the NCBI SRA database under the project name: BioProject ID: PRJNA758802.

Conflicts of Interest: The authors declare no conflict of interest. The funders had no role in the design of the study; in the collection, analyses, or interpretation of data; in the writing of the manuscript, or in the decision to publish the results.

References

1. REN21. Renewables 2020 Global Status Report. 2020. Available online: https://www.ren21.net/reports/global-status-report/?gclid=CjwKCAjwoP6LBhBlEiwAvCcthM9n4BM6oHikcrY_nWbzC811LiRCEiGVVkKEgzPT2NcaGYXiUiOBkhoCLR4QAvD_BwE (accessed on 1 November 2021).
2. Fawzy, S.; Osman, A.I.; Doran, J.; Rooney, D.W. Strategies for mitigation of climate change: A review. *Environ. Chem. Lett.* **2020**, *18*, 2069–2094. [CrossRef]
3. Østergaard, P.A. Comparing electricity, heat and biogas storages' impacts on renewable energy integration. *Energy* **2012**, *37*, 255–262. [CrossRef]
4. Lund, P.D.; Lindgren, J.; Mikkola, J.; Salpakari, J. Review of energy system flexibility measures to enable high levels of variable renewable electricity. *Renew. Sustain. Energy Rev.* **2015**, *45*, 785–807. [CrossRef]
5. Ketter, W.; Collins, J.; Saar-Tsechansky, M.; Marom, O. Information Systems for a Smart Electricity Grid. *ACM Trans. Manag. Inf. Syst.* **2018**, *9*, 1–22. [CrossRef]
6. Holladay, J.D.; Hu, J.; King, D.L.; Wang, Y. An overview of hydrogen production technologies. *Catal. Today* **2009**, *139*, 244–260. [CrossRef]
7. Thema, M.; Bauer, F.; Sterner, M. Power-to-Gas: Electrolysis and methanation status review. *Renew. Sustain. Energy Rev.* **2019**, *112*, 775–787. [CrossRef]
8. Sharma, S.; Ghoshal, S.K. Hydrogen the future transportation fuel: From production to applications. *Renew. Sustain. Energy Rev.* **2015**, *43*, 1151–1158. [CrossRef]
9. Herzog, A.; Tatsutani, M. A hydrogen future? An Economic and Environmental Assessment of Hydrogen Production Pathways. *Nat. Resour. Def. Counc.* **2005**, *23*.
10. Viswanathan, B. Hydrogen Storage. *Energy Sources* **2017**, 185–212. [CrossRef]
11. Andrei, H.; Badea, C.A.; Andrei, P.; Spertino, F. Energetic-Environmental-Economic Feasibility and Impact. *Energies* **2020**, *14*, 100. [CrossRef]
12. Campana, P.E.; Mainardis, M.; Moretti, A.; Cottes, M. 100% renewable wastewater treatment plants: Techno-economic assessment using a modelling and optimization approach. *Energy Convers. Manag.* **2021**, *239*, 114214. [CrossRef]
13. Luo, G.; Johansson, S.; Boe, K.; Xie, L.; Zhou, Q.; Angelidaki, I. Simultaneous hydrogen utilization and in situ biogas upgrading in an anaerobic reactor. *Biotechnol. Bioeng.* **2012**, *109*, 1088–1094. [CrossRef] [PubMed]
14. Global Methane Initiative. *Global Methane Emissions and Mitigation Opportunities*; Global Methane Initiative: Washington, DC, USA, 2010; Volume 2020, pp. 1–4. Available online: https://www.globalmethane.org/documents/gmi-mitigation-factsheet.pdf (accessed on 1 November 2021).
15. Angelidaki, I.; Treu, L.; Tsapekos, P.; Luo, G.; Campanaro, S.; Wenzel, H.; Kougias, P.G. Biogas upgrading and utilization: Current status and perspectives. *Biotechnol. Adv.* **2018**, *36*, 452–466. [CrossRef]
16. Piechota, G. Biogas/Biomethane Quality and Requirements for Combined Heat and Power (CHP) Units/Gas Grids with a Special Focus on Siloxanes-a Short Review. *Sustain. Chem. Eng.* **2022**, *3*, 1–10.
17. Piechota, G. Multi-step biogas quality improving by adsorptive packed column system as application to biomethane upgrading. *J. Environ. Chem. Eng.* **2021**, *9*, 105944. [CrossRef]
18. Angelidaki, I.; Karakashev, D.; Batstone, D.J.; Plugge, C.M.; Stams, A.J.M. *Biomethanation and Its Potential*, 1st ed.; Elsevier Inc.: Amsterdam, The Netherlands, 2011. [CrossRef]
19. Götz, M.; Lefebvre, J.; Mörs, F.; Koch, A.M.; Graf, F.; Bajohr, S.; Reimert, R.; Kolb, T. Renewable Power-to-Gas: A technological and economic review. *Renew. Energy* **2016**, *85*, 1371–1390. [CrossRef]
20. Rafrafi, Y.; Laguillaumie, L.; Dumas, C. Biological Methanation of H2 and CO2 with Mixed Cultures: Current Advances, Hurdles and Challenges. *Waste Biomass Valorization* **2020**, *12*, 5259–5282. [CrossRef]
21. Wahid, R.; Horn, S.J. Impact of operational conditions on methane yield and microbial community composition during biological methanation in in situ and hybrid reactor systems. *Biotechnol. Biofuels* **2021**, *14*, 170. [CrossRef]
22. Adnan, A.I.; Ong, M.Y.; Nomanbhay, S.; Chew, K.W. Technologies for Biogas Upgrading to Biomethane: A Review. *Bioengineering* **2019**, *6*, 92. [CrossRef] [PubMed]
23. Rusmanis, D.; O'Shea, R.; Wall, D.M.; Murphy, J.D. Biological hydrogen methanation systems–an overview of design and efficiency. *Bioengineered* **2019**, *10*, 604–634. [CrossRef]
24. Rittmann, S.; Seifert, A.; Herwig, C. Essential prerequisites for successful bioprocess development of biological CH4 production from CO2 and H2. *Crit. Rev. Biotechnol.* **2015**, *35*, 141–151. [CrossRef] [PubMed]
25. Peillex, J.P.; Fardeau, M.L.; Boussand, R.; Navarro, J.M.; Belaich, J.P. Growth of Methanococcus thermolithotrophicus in batch and continuous culture on H2 and CO2: Influence of agitation. *Appl. Microbiol. Biotechnol.* **1988**, *29*, 560–564. [CrossRef]
26. de Poorter, L.M.I.; Geerts, W.J.; Keltjens, J.T. Coupling of Methanothermobacter thermautotrophicus methane formation and growth in fed-batch and continuous cultures under different H2 gassing regimens. *Appl. Environ. Microbiol.* **2007**, *73*, 740–749. [CrossRef]
27. Strevett, K.A.; Vieth, R.F.; Grasso, D. Chemo-autotrophic biogas purification for methane enrichment: Mechanism and kinetics. *Chem. Eng. J. Biochem. Eng. J.* **1995**, *58*, 71–79. [CrossRef]
28. Szuhaj, M.; Ács, N.; Tengölics, R.; Bodor, A.; Rákhely, G.; Kovács, K.L.; Bagi, Z. Conversion of H2 and CO2 to CH4 and acetate in fed-batch biogas reactors by mixed biogas community: A novel route for the power-to-gas concept. *Biotechnol. Biofuels* **2016**, *9*, 102. [CrossRef]

29. Bassani, I.; Kougias, P.G.; Treu, L.; Angelidaki, I. Biogas Upgrading via Hydrogenotrophic Methanogenesis in Two-Stage Continuous Stirred Tank Reactors at Mesophilic and Thermophilic Conditions. *Environ. Sci. Technol.* **2015**, *49*, 12585–12593. [CrossRef]
30. Bassani, I.; Kougias, P.G.; Treu, L.; Porté, H.; Campanaro, S.; Angelidaki, I. Optimization of hydrogen dispersion in thermophilic up-flow reactors for ex situ biogas upgrading. *Bioresour. Technol.* **2017**, *234*, 310–319. [CrossRef] [PubMed]
31. Agneessens, L.M.; Ottosen, L.D.M.; Voigt, N.V.; Nielsen, J.L.; de Jonge, N.; Fischer, C.H.; Kofoed, M.V.W. In-situ biogas upgrading with pulse H2additions: The relevance of methanogen adaption and inorganic carbon level. *Bioresour. Technol.* **2017**, *233*, 256–263. [CrossRef]
32. Luo, G.; Angelidaki, I. Integrated biogas upgrading and hydrogen utilization in an anaerobic reactor containing enriched hydrogenotrophic methanogenic culture. *Biotechnol. Bioeng.* **2012**, *109*, 2729–2736. [CrossRef]
33. Strübing, D.; Huber, B.; Lebuhn, M.; Drewes, J.E. High performance biological methanation in thermophilic anaerobic trickle bed reactors. *Bioresour. Technol.* **2017**, *245*, 1176–1183. [CrossRef] [PubMed]
34. Rachbauer, L.; Beyer, R.; Bochmann, G.; Fuchs, W. Characteristics of adapted hydrogenotrophic community during biomethanation. *Sci. Total Environ.* **2017**, *595*, 912–919. [CrossRef] [PubMed]
35. Ács, N.; Szuhaj, M.; Wirth, R.; Bagi, Z.; Maróti, G.; Rákhely, G.; Kovács, K.L. Microbial Community Rearrangements in Power-to-Biomethane Reactors Employing Mesophilic Biogas Digestate. *Front. Energy Res.* **2019**, *7*, 132. [CrossRef]
36. Bassani, I.; Kougias, P.G.; Angelidaki, I. In-situ biogas upgrading in thermophilic granular UASB reactor: Key factors affecting the hydrogen mass transfer rate. *Bioresour. Technol.* **2016**, *221*, 485–491. [CrossRef] [PubMed]
37. Afgan, E.; Baker, D.; van den Beek, M.; Blankenberg, D.; Bouvier, D.; Čech, M.; Chilton, J.; Clements, D.; Coraor, N.; Eberhard, C.; et al. The Galaxy platform for accessible, reproducible and collaborative biomedical analyses: 2016 update. *Nucleic Acids Res.* **2016**, *44*, W3–W10. [CrossRef]
38. Schmieder, R.; Edwards, R. Quality control and preprocessing of metagenomic datasets. *Bioinformatics* **2011**, *27*, 863–864. [CrossRef]
39. Menzel, P.; Ng, K.L.; Krogh, A. Fast and sensitive taxonomic classification for metagenomics with Kaiju. *Nat. Commun.* **2016**, *7*, 11257. [CrossRef]
40. Mende, D.R.; Letunic, I.; Huerta-Cepas, J.; Li, S.S.; Forslund, K.; Sunagawa, S.; Bork, P. ProGenomes: A resource for consistent functional and taxonomic annotations of prokaryotic genomes. *Nucleic Acids Res.* **2017**, *45*, D529–D534. [CrossRef]
41. Letunic, I.; Bork, P. Interactive Tree of Life (iTOL) v4: Recent updates and new developments. *Nucleic Acids Res.* **2019**, *47*, 256–259. [CrossRef]
42. Siriwongrungson, V.; Zeng, R.J.; Angelidaki, I. Homoacetogenesis as the alternative pathway for H2 sink during thermophilic anaerobic degradation of butyrate under suppressed methanogenesis. *Water Res.* **2007**, *41*, 4204–4210. [CrossRef]
43. Bouanane-Darenfed, A.; Hania, W.B.; Cayol, J.L.; Ollivier, B.; Fardeau, M.L. Reclassification of acetomicrobium faecale as caldicoprobacter faecalis comb. Nov. *Int. J. Syst. Evol. Microbiol.* **2015**, *65*, 3286–3288. [CrossRef]
44. Avdiyuk, K.V. Keratinolytic Enzymes: Producers, Physical and Chemical Properties. Application for Biotechnology. *Biotechnol. Acta* **2019**, *12*, 27–45. [CrossRef]
45. Fortina, M.G.; Pukall, R.; Schumann, P.; Mora, D.; Parini, C.; Manachini, P.L.; Stackebrandt, E. Ureibacillus gen. nov., a new genus to accommodate Bacillus thermosphaericus (Andersson et al. 1995), emendation of Ureibacillus thermosphaericus and description of Ureibacillus terrenus sp. nov. *Int. J. Syst. Evol. Microbiol.* **2001**, *51*, 447–455. [CrossRef]
46. Weon, H.Y.; Lee, S.Y.; Kim, B.Y.; Noh, H.J.; Schumann, P.; Kim, J.S.; Kwon, S.W. Ureibacillus composti sp. nov. and Ureibacillus thermophilus sp. nov., isolated from livestock-manure composts. *Int. J. Syst. Evol. Microbiol.* **2007**, *57*, 2908–2911. [CrossRef]
47. Maus, I.; Koeck, D.E.; Cibis, K.G.; Hahnke, S.; Kim, Y.S.; Langer, T.; Kreubel, J.; Erhard, M.; Bremges, A.; Off, S.; et al. Unraveling the microbiome of a thermophilic biogas plant by metagenome and metatranscriptome analysis complemented by characterization of bacterial and archaeal isolates. *Biotechnol. Biofuels* **2016**, *9*, 1–28. [CrossRef]
48. Shiratori, H.; Ohiwa, H.; Ikeno, H.; Ayame, S.; Kataoka, N.; Miya, A.; Beppu, T.; Ueda, K. Lutispora thermophila gen. nov., sp. nov., a thermophilic, spore-forming bacterium isolated from a thermophilic methanogenic bioreactor digesting municipal solid wastes. *Int. J. Syst. Evol. Microbiol.* **2008**, *58*, 964–969. [CrossRef]
49. Logroño, W.; Popp, D.; Kleinsteuber, S.; Sträuber, H.; Harms, H.; Nikolausz, M. Microbial resource management for ex situ biomethanation of hydrogen at alkaline ph. *Microorganisms* **2020**, *8*, 614. [CrossRef] [PubMed]
50. Wasserfallen, A.; Nölling, J.; Pfister, P.; Reeve, J.; de Macario, E.C. Phylogenetic analysis of 18 thermophilic Methanobacterium isolates supports the proposals to create a new genus, Methanothermobacter gen. nov., and to reclassify several isolates in three species, Methanothermobacter thermautotrophicus comb. nov., Methano. *Int. J. Syst. Evol. Microbiol.* **2000**, *50*, 43–53. [CrossRef] [PubMed]
51. Cao, Y.; Wang, L.; Qian, Y.; Xu, Y.; Wu, H.; Zhang, J.; Huang, H.; Chang, Z. Contributions of thermotolerant bacteria to organic matter degradation under a hyperthermophilic pretreatment process during chicken manure composting. *BioResources* **2019**, *14*, 6747–6766. [CrossRef]
52. Gorliczay, E.; Boczonádi, I.; Kiss, N.É.; Tóth, F.A.; Pabar, S.A.; Biró, B.; Kovács, L.R.; Tamás, J. Microbiological effectivity evaluation of new poultry farming organic waste recycling. *Agric* **2021**, *11*, 683. [CrossRef]
53. Vaz-Moreira, I.; Faria, C.; Nobre, M.F.; Schumann, P.; Nunes, O.C.; Manaia, C.M. Paenibacillus humicus sp. nov., isolated from poultry litter compost. *Int. J. Syst. Evol. Microbiol.* **2007**, *57*, 2267–2271. [CrossRef]
54. McInerney, M.J.; Bryant, M.P.; Hespell, R.B.; Costerton, J.W. Syntrophomonas wolfei gen. nov. sp. nov., an anaerobic, syntrophic, fatty acid-oxidizing bacterium. *Appl. Environ. Microbiol.* **1981**, *41*, 1029–1039. [CrossRef]

55. Westerholm, M.; Roos, S.; Schnürer, A. Syntrophaceticus schinkiigen. nov., sp. nov., an anaerobic, syntrophic acetate-oxidizing bacterium isolated from a mesophilic anaerobic filter. *FEMS Microbiol. Lett.* **2010**, *309*, 100–104. [CrossRef]
56. United Nations. Adoption of the Paris Agreement, Proposal by the President, Draft decision. In Proceedings of the Twenty-First Session of the Conference of the Parties (COP 21), Paris, France, 11 December 2015; Volume 32, p. 21932. Available online: http://unfccc.int/resource/docs/2015/cop21/eng/l09r01.pdf (accessed on 1 November 2021).
57. Csedő, Z.; Zavarkó, M.; Vaszkun, B.; Koczkás, S. Hydrogen Economy Development Opportunities by Inter-Organizational Digital Knowledge Networks. *Sustainability* **2021**, *13*, 9194. [CrossRef]
58. Kakuk, B.; Wirth, R.; Maróti, G.; Szuhaj, M.; Rakhely, G.; Laczi, K.; Kovács, K.L.; Bagi, Z. Early response of methanogenic archaea to H2 as evaluated by metagenomics and metatranscriptomics. *Microb. Cell Fact.* **2021**, *20*, 127. [CrossRef] [PubMed]
59. Kleinsteuber, S. Metagenomics of Methanogenic Communities in Anaerobic Digesters. In *Biogenesis of Hydrocarbons Handbook of Hydrocarbon and Lipid Microbiology*; Springer: Cham, Switzerland, 2019; pp. 337–359. [CrossRef]
60. Corbellini, V.; Kougias, P.G.; Treu, L.; Bassani, I.; Malpei, F.; Angelidaki, I. Hybrid biogas upgrading in a two-stage thermophilic reactor. *Energy Convers. Manag.* **2018**, *168*, 1–10. [CrossRef]
61. Mulat, D.G.; Mosbæk, F.; Ward, A.J.; Polag, D.; Greule, M.; Keppler, F.; Nielsen, J.L.; Feilberg, A. Exogenous addition of H2 for an in situ biogas upgrading through biological reduction of carbon dioxide into methane. *Waste Manag.* **2017**, *68*, 146–156. [CrossRef] [PubMed]
62. Bagi, Z.; Ács, N.; Böjti, T.; Kakuk, B.; Rákhely, G.; Strang, O.; Szuhaj, M.; Wirth, R.; Kovács, K.L. Biomethane: The energy storage, platform chemical and greenhouse gas mitigation target. *Anaerobe* **2017**, *46*, 13–22. [CrossRef]
63. Agneessens, L.M.; Ottosen, L.D.M.; Andersen, M.; Olesen, C.B.; Feilberg, A.; Kofoed, M.V.W. Parameters affecting acetate concentrations during in-situ biological hydrogen methanation. *Bioresour. Technol.* **2018**, *258*, 33–40. [CrossRef]
64. Waldron, C.R.; Becker-Vallone, C.A.; Eveleigh, D.E. Isolation and characterization of a cellulolytic actinomycete Microbispora bispora. *Appl. Microbiol. Biotechnol.* **1986**, *24*, 477–486. [CrossRef]
65. Li, Y.; Chen, F.; Dong, K.; Wang, G. Actinotalea ferrariae sp. nov., isolated from an iron mine, and emended description of the genus Actinotalea. *Int. J. Syst. Evol. Microbiol.* **2013**, *63*, 3398–3403. [CrossRef]
66. Stackebrandt, E.; Kandler, O. Taxonomy of the genus Cellulomonas, based on phenotypic characters and deoxyribonucleic acid-deoxyribonucleic acid homology, and proposal of seven neotype strains. *Int. J. Syst. Bacteriol.* **1979**, *29*, 273–282. [CrossRef]
67. Kushkevych, I.; Cejnar, J.; Vítězová, M.; Vítěz, T.; Dordević, D.; Bomble, Y. Occurrence of thermophilic microorganisms in different full scale biogas plants. *Int. J. Mol. Sci.* **2020**, *21*, 283. [CrossRef]
68. Nouioui, I.; Carro, L.; García-López, M.; Meier-Kolthoff, J.P.; Woyke, T.; Kyrpides, N.C.; Pukall, R.; Klenk, H.P.; Goodfellow, M.; Göker, M. Genome-based taxonomic classification of the phylum actinobacteria. *Front. Microbiol.* **2018**, *9*, 1–119. [CrossRef]
69. Hattori, S.; Kamagata, Y.; Hanada, S. A Strictly Anaerobic, Thermophilic, Syntrophic Acetate-Oxidizing Bacterium. *Int. J. Syst. Evol. Microbiol.* **2000**, *50*, 1601–1609. [CrossRef] [PubMed]
70. Sattar, A.; Zakaria, Z.; Abu, J.; Aziz, S.A.; Rojas-Ponce, G. Isolation of Mycobacterium avium and other nontuberculous mycobacteria in chickens and captive birds in peninsular Malaysia. *BMC Vet. Res.* **2021**, *17*, 13. [CrossRef] [PubMed]
71. Dyksma, S.; Jansen, L.; Gallert, C. Syntrophic acetate oxidation replaces acetoclastic methanogenesis during thermophilic digestion of biowaste. *Microbiome* **2020**, *8*, 105. [CrossRef] [PubMed]
72. FitzGerald, J.A.; Allen, E.; Wall, D.M.; Jackson, S.A.; Murphy, J.D.; Dobson, A.D.W. Methanosarcina play an important role in anaerobic co-digestion of the seaweed ulva lactuca: Taxonomy and predicted metabolism of functional microbial communities. *PLoS ONE* **2015**, *10*, e0142603. [CrossRef] [PubMed]
73. Silva, T.C.D.; Isha, A.; Chandra, R.; Vijay, V.K.; Subbarao, P.M.V.; Kumar, R.; Chaudhary, V.P.; Singh, H.; Khan, A.A.; Tyagi, V.K.; et al. Enhancing methane production in anaerobic digestion through hydrogen assisted pathways—A state-of-the-art review. *Renew. Sustain. Energy Rev.* **2021**, *151*, 111536. [CrossRef]

MDPI
St. Alban-Anlage 66
4052 Basel
Switzerland
Tel. +41 61 683 77 34
Fax +41 61 302 89 18
www.mdpi.com

Energies Editorial Office
E-mail: energies@mdpi.com
www.mdpi.com/journal/energies

www.ingramcontent.com/pod-product-compliance
Lightning Source LLC
LaVergne TN
LVHW070607100526
838202LV00012B/590